A Catalogue of Manuscripts and Drawings in the General Library of The Natural History Museum, London

The Natural History Museum

A Catalogue of Manuscripts and Drawings in the General Library of The Natural History Museum, London

John C. Thackray

MANSELL

Historical Studies in the Life and Earth Sciences No. 4

First published 1995 by
Mansell Publishing Limited, *A Cassell imprint*
Wellington House, 125 Strand, London WC2R 0BB, England
215 Park Avenue South, New York, NY 10003, USA

British Library Cataloguing-in-Publication Data
Catalogue of Manuscripts and Drawings in
the General Library of The Natural
History Museum. – (Historical Studies in
the Life and Earth Sciences; No. 4)
I. Thackray, John C. II. Series
016.508

ISBN 0-7201-2291-0

Library of Congress Cataloging-in-Publication Data
Natural History Museum (London). General Library.
A catalogue of manuscripts and drawings in the General Library of The Natural History Museum / John C. Thackray.
p. cm.—(Historical studies in the life and earth sciences : no. 4)
Includes bibliographical references and indexes.
ISBN 0-7201-2291-0
1. Natural history—Manuscripts—Catalogs. 2. Natural history Illustration—Catalogs. 3. Natural history—History—Sources—Bibliography—Catalogs. 4. Manuscripts—England—London—Catalogs. 5. Natural History Museum (London, England). General Library—Catalogs. I. Thackray, John C. II. Title. III. Series.
Z7409.N275 1995
[QH70.A1]
016.508 - dc20 95-6190
CIP

Typeset in 11/16 pt Palatino by Colset Pte Ltd, Singapore

Printed and bound in Great Britain

Contents

Acknowledgements

I am grateful to my colleagues in the Department of Library and Information Services for their ready responses to my constant enquiries over the two years of compilation of the catalogue. In particular I thank Mrs Ann Datta who contributed catalogue entries for the Loten, Martens, 'Port Jackson Painter', Sarah Stone and Alfred Waterhouse collections of drawings. I have profited from advice from members of the scientific departments of the Museum, in particular Mr R. J. Cleevely (Palaeontology), and Miss D. M. Hills, Dr P. B. Morden and Ms K. Way (Zoology). Outside the Museum, Dr D. E. Allen, Mr J. H. Appleby, Mr I. Bailey, Mr S. P. Dance and Mrs C. E. Jackson all responded kindly and promptly to calls for information.

Introduction

The Natural History Museum traces its history back to 1753, the year in which the British Museum was founded by Act of Parliament. From this date until 1880 the natural history collections of the Museum were displayed alongside the historical and artistic materials in a single building in Bloomsbury. The movement to house the natural history collections in a separate building began with the appointment of Richard Owen (1804–1892) as Superintendent of the natural history departments in 1856. He, together with J. E. Gray (1800–1875) and the other keepers, campaigned tirelessly for adequate space in a new building (Rupke, 1994, chapter 1). The new museum in South Kensington was designed by Alfred Waterhouse, and was ready for occupation in 1880 (Girouard, 1981).

During the years in Bloomsbury, the primary source of literature for the scientists was the great British Museum Library, cared for by the Department of Printed Books. The Principal Librarian had no intention of allowing his Library to be denuded of natural history books, and so the British Museum Act 1878, allowing for the transfer of the natural history departments and their collections, barred the transfer of all but duplicate books from the Library. For this reason, when the departments of Botany, Geology, Mineralogy and Zoology moved over from Bloomsbury with their vast scientific collections, they brought with them only the few thousand books, manuscripts and drawings that had been purchased departmentally over the years. These departmental libraries received generous purchase grants through the 1880s and 1890s to build up their size and scope.

The Trustees accepted that the new museum would need a central collection of those books and periodicals that were not specific to one subject. It was named the General Library, and was placed under the supervision of a committee of the four keepers. Bernard Barham Woodward (1853–1930) was transferred from Bloomsbury to be Assistant in Charge of the Library, a post that he held until 1920. The Director took over supervision of the General Library in 1884, and it remained part of the Directorate until the Department of Library Services was set up in 1975 (Stearn, 1981, chapter 17). Waterhouse's plans for the Museum made no provision for this central collection, and the General Library had to be housed in a corridor on the east side of the Central Hall until 1959, when a northern extension to the original building was opened.

Woodward enlisted the help of the keepers and other staff to build up the General Library collections, and by the end of 1884 he had 12,000 volumes in stock. By 1900 the figure was 19,000 volumes and 5,500 sheets of maps (Sawyer, 1971, p. 82).

Only a small number of manuscripts and drawings were acquired in the early years, mostly as gifts. These included the watercolour drawings of Thomas Baines (1886), watercolour drawings by Chinese artists (1894) and the volume of pencil and pen and ink drawings of George Shaw (1896). By far the largest of the early acquisitions was the Richard Owen Collection. This was presented in 1893, 1908 and 1915 by the trustees of Owen's will, Charles Davies Sherborn and Miss Emily Owen respectively, and includes manuscripts, drawings, annotated books, notebooks and correspondence of Owen himself, William Clift (1775–1849) and Sir Everard Home (1756–1832). It remains an important source, not only for historians, but for the Museum's collection managers, who are concerned with the documentation of historic specimens. The thirty-two volumes of the Owen–Clift correspondence is the single most heavily used collection in the General Library Manuscripts (Gruber 1992).

C. D. Sherborn (1861–1942), a bibliographer who worked as a researcher in the libraries for many years, was the founder of the Handwriting Collection. This great resource contains a representative sample of the handwriting of several thousand naturalists of all disciplines, nationalities and dates. The collection was in the course of formation by 1904, was largely completed by the end of the 1920s, and still receives occasional additions. The aim of the collection was to provide a means of identifying unsigned handwriting, particularly on specimen labels. Almost by chance, many of the letters it includes are themselves of great historical interest.

The General Library Manuscript collection grew only slowly through the middle years of this century, and was overshadowed by the manuscripts and drawings in the departmental libraries. However, the position changed in 1969 with the purchase of the first portions of both the Sowerby and Gunther collections. The archive of the Sowerby family of natural history illustrators and publishers was acquired between 1969 and 1983, and is rich in early nineteenth-century botany, conchology and palaeontology. The Gunther papers on the other hand, acquired by gift and purchase between 1969 and 1980, principally complement the Museum Archives in documenting the zoological work of Albert Gunther, Keeper of Zoology, between about 1875 and 1910.

Large and important collections have continued to be acquired, with, among others, the letters from T. A. Knight to Sir Joseph Banks (1973), the Muriel Dawson drawings (1975), the Warren Dawson correspondence (1978), the Ray Society Archive (1985), the papers of G. C. Wallich (1985–1986) and those of Sir Maurice Yonge (1990).

These accessions have transformed a modest and somewhat unfocused collection into a major resource of international importance, which is consulted by scholars throughout the world. The General Library Manuscripts now comprise 381 collections under 129 author/institutional headings, and occupy 110 metres of shelving. Among the strengths of the collection are the holdings relating to Africa (Baines, Collier, Rosevear, Scott and Sibree), to India (Ditmas, Hume, Sykes and Tonge), and to Australia (Martens, the 'Port Jackson Painter', Wilkins and Yonge). Important material relating to British natural history is to be found in the collections of Adams, Muriel Dawson, the Sale Natural History Society, Saunders, Thompson, Turner, Victoria County History, Wintle and B. B. Woodward. Collections of photographs and films are to be found among the manuscripts and drawings, notably those of A. H. Bastin, De Winton, Frances Pitt and Sir Hubert Wilkins. Although the bulk of the collection is nineteenth and early twentieth century, there are earlier items to be found under the names of Bartalini, Fischer, Menzies, Montagu, Pulteney and Richardson.

It is hoped that the provision of this, its first complete catalogue, will increase the knowledge, and use made, of the collection.

Arrangement

Collections are listed under the name of the individual author, artist, compiler or recipient, the corporate body responsible, the name of a ship or, failing all else, the country of origin. Each entry consists of the following elements:

Heading.
Biographical or historical notes.
Biographical or historical reference.
Collection entries:
  physical description;
  title or description;
  date;
  further explanatory notes, including details of major correspondents in a letter collection, with the number of letters and date span;
  finding aid note;
  provenance;
  shelf mark.

The following abbreviations are used throughout the catalogue:
DNB – Dictionary of National Biography (edited by L. Stephen, et al., 1885–1990, 65 volumes and later supplements, London).
DSB – Dictionary of Scientific Biography (edited by C. C. Gillespie, 1970–1980, 16 volumes, New York).
Ms – manuscript
Mss – manuscripts
Ts – typescript
Tss – typescripts.

A Note on Names

'The Natural History Museum' is used to refer to the Museum in South Kensington from 1881 to the present even though, for most of that period, its official name was 'British Museum (Natural History)'. 'The British Museum' is used for the period prior to 1881, when the natural history departments were located in Bloomsbury.

Bibliography

Girouard, M. 1981. *Alfred Waterhouse and The Natural History Museum.* 64pp. London: The Natural History Museum.

Gruber, J. W. 1992. The Richard Owen correspondence: an introductory essay, in Gruber, J. W. and Thackray, J. C. 1992. *Richard Owen Commemoration.* 181pp. London: Natural History Museum Publications.

Rupke, N. 1994. *Richard Owen, Victorian Naturalist.* 462pp. New Haven, CT: Yale University Press.

Sawyer, F. C. 1971. A short history of the libraries and list of Mss and original drawings in The Natural History Museum. *Bulletin of the British Museum (Natural History),* Historical Series 4(2):79–291.

Stearn, W. T. 1981. *The Natural History Museum at South Kensington.* 414pp. London: Heinemann.

The Catalogue

1. ADAMS, Lionel Ernest (1854–1945)

Born in London, Adams taught in schools in Manchester, Stafford and Penistone, and spent his retirement on the Isle of Wight. He was interested in a wide range of natural history subjects, but specialized in the British land and freshwater Mollusca. His most important publication was *The Collector's Manual of British Land and Freshwater Shells* (1885, 2nd edn. 1896). Adams was President of the Conchological Society from 1888 to 1890. He gave his collection of shells to the Conchological Society in 1940, and it passed to the Museum in 1963. A Ms catalogue of his collection is held in the Zoology Library Mollusca Manuscripts.

Biographical Reference: 1946. *Journal of Conchology*, 22:171–172.

Collection

(1) One letterbook: some correspondence of L. E. Adams, 1873–1942. The collection contains 90 letters to L. E. Adams and one to his father, Henry Adams, from 73 correspondents, and also 48 signatures.
L MSS ADA

2. AGASSIZ, Jean Louis Rodolphe (1807–1873)

Agassiz was born and grew up in Switzerland, attending universities in Zurich, Heidelberg and Munich. He was a professor at the College of Neufchâtel from 1832 to 1847, after which he emigrated to the United States to become

Professor of Zoology and Geology at Harvard University, where he remained until his death. Agassiz's most important works are his *Recherches sur les Poissons Fossiles* (1833–1843) and *Études sur les Glaciers* (1840).

Biographical Reference: *DSB*.

Collection

(1) Two Ms notebooks: containing notes on anatomy, chemistry and physiology taken down by Agassiz when he was a student in Zurich, 1824–1825. Presented by A. H. Higginson, Agassiz's grandson, in 1951.
L MSS AGA A

3. **ALDER, Joshua** (1792–1867)

Alder was born, educated, and in business in Newcastle upon Tyne. He made a study of the marine zoology of the seas around northern England and Scotland, specializing in molluscs and zoophytes. His most important publication, produced in collaboration with Albany Hancock, was *Monograph of the British Nudibranchiate Mollusca* (1845–1855). Alder was a founder member of both the Natural History Society of Northumberland, Durham and Newcastle Upon Tyne (1829) and the Tyneside Naturalists' Fieldclub (1846).

Biographical Reference: *DNB*.

Collection

(1) Seven letterbooks: the correspondence of Alder and Revd Alfred Merle Norman, 1826–1911. There are c.1000 letters from 300 correspondents, including J. Alder (93, 1856–1866), E. Forbes (33, 1836–1853), A. Hancock (14, 1842–1869), T. Hincks (17, 1855–1877), J. G. Jeffreys (93, 1832–1877), A. M. Norman (64, 1859–1903), C. W. Peach (12, 1847–1868), D. Robertson (31, 1864–1876), T. Scott (39, 1892–1906) and W. Thompson (30, 1834–1847). The letters are mostly concerned with the marine zoology around Scotland and the north of England. There is a catalogue of the collection by A. F. Wood, 1963. The collection was presented by Mrs E. C. L. Norman in 1937, and was arranged and annotated by C. D. Sherborn.
L MSS ALD

4. ALLEN, David Elliston (b.1932)

David Allen was educated at Rugby School, where he developed an interest in field botany, and at Cambridge University, where he studied archaeology and anthropology. His career has been in grants administration, first for the Economic and Social Science Research Council (1967–1986), and for the Wellcome Trust, in conjunction with research and teaching at the Wellcome Institute for the History of Medicine. He was President of the Society for the Bibliography of Natural History from 1978 to 1981, and of the Botanical Society of the British Isles from 1985 to 1987. Dr Allen is author of books and papers on the history of natural history, including *The Naturalist in Britain: a Social History* (1976).

Biographical Reference: personal information.

Collection

(1) Three boxes of letters: D. E. Allen's correspondence on the history of natural history, 1957–1987. Letters are arranged chronologically. Correspondence less than fifteen years old is not available for study. Presented by Dr Allen from 1985.
L MSS ALL

5. APPLEBY, John Hilary (b.1938)

John Appleby is a writer, researcher and translator, and a member of the Society of Authors and the Translation Association. He has made a particular study of Anglo-Russian history.

Biographical Reference: personal information.

Collection

(1) One Ts in a folder: a transcript of part of the 'Kamerny Katalog' of books (1742), held in the Library of the Russian Academy of Sciences, St Petersburg, c.1980. The books relate to medicine and natural history. Presented by the author, 1981.
L MSS APP

6. AUDOUIN, Jean Victor (1797–1841)

Audouin studied in Rheims and Paris before becoming secretary to the mineralogist, Alexandre Brongniart (1770–1847). He worked at the Muséum National d'Histoire Naturelle, first as an assistant to Lamarck and Latreille, and, from 1833, as Professor of Zoology. Audouin studied marine invertebrates along the north coast of France, before turning to agricultural entomology in the 1830s. He collaborated with H. Milne Edwards (1800–1885) to produce *Recherches pour Servir à l'Histoire Naturelle du Littoral de la France* (1832–1834). He was one of the founders of the Société Entomologique de France in 1833.

Biographical Reference: *DSB*.

Collection

(1) One bound Ms: [*Arachnida*], c.1834, with pen-and-ink original drawings. In French; missing pages 1–2. Published in English in *The Cyclopaedia of Anatomy and Physiology* (edited by R. B. Todd, 1836). Presented by C. D. Sherborn, 1908, formerly in the collection of Sir Richard Owen.
L MSS AUD

7. BAINES, Thomas (1822–1875)

Baines, artist and explorer, was born in Norfolk, and began to draw and paint at an early age. He travelled in South Africa from 1842 to 1854, in northern Australia from 1855 to 1856, in central Africa with Livingstone from 1858 to 1861, and in southern Africa from 1868 until his death. Among his travel books is *The Gold Regions of South Eastern Africa* (1877). Baines was made a Fellow of the Royal Geographical Society in 1857.

Biographical Reference: *DNB*.

Collections

(1) One volume of drawings: *Original Sketches in Watercolour and Pencil of Scenes, Animals etc., Made in South-east Africa*, 1859; 1869–1871. Two pain-

ings of mangrove swamps were completed with Livingstone in 1859, the remaining 153 date from Baines's work for the South African Gold Fields Exploration Company, described in *The Gold Regions of South Eastern Africa* (1877). Purchased 1886.
L MSS BAI A

(2) One folder of Mss: four pencil drawings, three of them coloured, of zebra and quagga, with two leaves of notes, 1862–1864. Drawings were made while Baines was on James Chapman's travels to the Victoria Falls, and are cited in Baines's *Explorations in South-West Africa* (1864). One is reproduced in *The Gold Regions . . .*, p. 71. Bequeathed by Lord Rothschild, 1937.
L MSS BAI B

8. **BANKS, Sir Joseph (1743–1820)**

Banks was a member of a wealthy family of Lincolnshire landowners. He was educated at Harrow, Eton and Oxford, and became interested in botany. Banks visited Labrador and Newfoundland in 1766, Iceland in 1772, and travelled round the world on Cook's first voyage from 1768 to 1771. He spent the rest of his life in London, building up his library and collections, and becoming the most important patron and adviser in science and natural history of the period. Banks was President of the Royal Society from 1778 to 1820. His library and herbarium were bequeathed to Robert Brown (1773–1858) on his death, and passed eventually to the British Museum.

Biographical Reference: *DSB*.

Collections

(1) One Ms leaf: letter to the Rt. Hon. Charles Greville, Sept. 1802. Purchased 1977.
L MSS BAN A

(2) One Ms leaf: letter to Charles James, 21 June 1811. Purchased 1978.
L MSS BAN A

(3) One Ms leaf: letter to an unknown correspondent, 9 April 1805. Purchased 1981.
L MSS BAN A

(4) One folder of Mss: seven letters to Banks from A. Moon and A. Wedderburn, and from Banks to W. Bligh, T. J. Bell, R. Chaplin, G. Filmer and A. Wedderburn, 1787–1819.
L MSS BAN A

(5) One folder of Mss: eight notes, lists, journal extracts and letters relating to Banks. Presented by C. Kirke Swann, 1937.
L MSS BAN A

9. **BARTALINI, Biagio (1746–1822)**

Professor of Natural History, Chemistry and Botany at Sienna, author of *Catalogo delle Piante che Nascono Spontaneamente. . .* (1776).

Biographical Reference: title page of his book.

Collection

(1) One Ms volume: *Analisi delle Sostanze Animali*, bound with *Cognizioni Prelimininari per Apprendere il Sistema delle Piante Secondo Linneo*, Sienna, 1814. Purchased 1973.
L MSS BAR

10. **BASTIN, Alfred Harold (1875–1962)**

Born in Torquay, Bastin set up as an entomological dealer in Reading in 1901, turning to photography and model-making in 1909. He worked in partnership with his brother Leonard for many years. He was the author of *Insects: their Life Histories and Habits* (1913).

Biographical Reference: personal information.

Collection

(1) Photograph collection: c.15,000 prints and glass negatives of natural history subjects, c.1910–1956, with four volumes of catalogues and indexes. Presented by Topham Picture Library, 1986.
L MSS BAS

11. **BATES, George Latimer (1863–1940)**

Bates was born in Illinois and attended colleges in Galesburg and Chicago. He taught in a boys school for a while but, being keen on travel and natural history, decided to try to earn his living as a collector. He lived in West Africa from 1895 until 1928, when he moved to England. He payed a final visit to West Africa in 1931, before turning his attention to Arabia, which he visited in 1934. Bates was the author of *Handbook of the Birds of West Africa* (1930). He worked in the Bird Room of The Natural History Museum for many years, and was a generous donor to the collections.

Biographical Reference: 1940. Obituary. *Ibis*, 14th series 4: 343–348.

Collection

(1) One folder of Mss: eleven letters to William E. De Winton, 1896–1900, from the French Congo, with notes and animal stories. Acquired from Mrs De Winton, 1968.
L MSS BAT

12. **BAYLES, Nathaniel (1717/18–1791)**

Bayles was a surgeon of Newcastle upon Tyne, and a friend of Thomas Bewick (1753–1828), the wood engraver. He patented an improved whaling harpoon, and became the official sword-bearer to the City of Newcastle.

Biographical Reference: Welford, R. 1895. *Men of Mark 'twixt Tyne and Tweed*. 3 volumes. London.

Collection

(1) One Ms volume: *Directions for Collecting and Preserving Natural Curiosities found in the North Seas and Greenland &c.*, 1785. Purchased 1980.
L MSS BAY

13. **BELT, Thomas (1832–1878)**

Born and educated in Newcastle upon Tyne, Belt travelled in Australia, Nova Scotia, Nicaragua and Siberia between the years 1852 and 1878. He worked as a mining geologist, and was particularly interested in glacial geology. He wrote *The Naturalist in Nicaragua* (1874), and was a Fellow of the Geological Society of London.

Biographical Reference: DNB.

Collection

(1) Two letterbooks: 180 letters to Belt, 1864–1878, from writers including H. W. Bates (14, 1871–1877), J. G. Michie (14, 1874–1878) and D. Sharp (14, 1873–1877). A finding aid is available. Presented October 1938.
L MSS BEL

14. **BLACKWALL, John (1790–1881)**

John Blackwall was born and educated in Manchester. He was in business with his father until 1833, when he moved to North Wales. Blackwall was interested in natural history from an early age, and published a number of papers on birds and spiders. He was the author of the Ray Society (q.v.) monograph *A History of the Spiders of Great Britain and Ireland* (1861–1864).

Biographical Reference: DNB.

Collection

(1) One Ms in a folder: *Report on the Structure and Uses of the Palpi of the Araneidea*, [1844], annotated by Richard Owen. Purchased 1970.
L MSS BLA

15. **BLOK, Arthur (1882–1974)**

Born and educated in London, Arthur Blok was an electrical engineer, a technical and administrative civil servant, academic and collector. He developed a great interest in shells, and built up a large and immaculate collection, largely by purchase, as well as a fine conchological library. Both his library and collection were presented to the Hebrew University in Jerusalem. Blok was a member of the Conchological Society, where he gave a number of notable lectures, and was an occasional contributor to the *Journal of Conchology* and *Proceedings of the Malacological Society*.

Biographical Reference: 1976. Obituary. *Journal of Conchology* 29:67–68.

Collections

(1) Two letterbooks: 650 letters of malacologists and other naturalists, c.1800–1960, addressed to A. Blok, S. P. Dance, O. A. L. Morch, A. J. Piele, J. R. Le B. Tomlin, R. Winckworth and others. The letters were assembled as an autograph collection by Arthur Blok and added to by S. Peter Dance. Acquired from S. Peter Dance, 1975.
L MSS BLO

(2) One scrapbook: *Conchyliorum*, an album containing drawings, prints and photographs of molluscs and their shells, portraits of malacologists, and postage stamps, 1761–1970. Included are watercolour drawings by Guy L. Wilkins and prints from Thomas Martyn's *Universal Conchologist* (1784–1792). Presented by Mrs H. Block, 1994.
L MSS BLO

16. **BLOXAM, Andrew (1801–1878)**

Born in Rugby, Bloxam accepted the post of naturalist on board the frigate HMS *Blonde* in 1824, and travelled with her to the Sandwich Islands, on a voyage lasting eighteen months. On his return Bloxam took holy orders and lived first at Twycross and later Harborough Magna, both in Leicestershire. His work is referred to in Lord Byron's *Voyage of HMS Blonde to the Sandwich Islands* (1826). Many of the natural history specimens collected on the voyage were presented to the British Museum. Presented by A. Roby Bloxam, 1909.

Biographical Reference: *DNB*.

Collection

(1) One box of Mss: notes made on the voyage of HMS *Blonde*, [1824–1826], with other letters relating to the collection. These include a rough diary of the voyage; a transcript of part of the diary; two notebooks of observations on birds, with sketches; rough notes on birds; descriptions of birds found in Chile; natural history notes made in Chile and on the Galápagos Islands; Bloxam's report on the natural history of the Sandwich Islands. Also included are letters from A. Newton and H. O. Forbes to A. Roby Bloxam about the collection (1890–1898), and letters from E. Hartert and C. E. Fagan about presentation of the collection to the Museum (1909).
L MSS BLOX

17. **BOLTON, W.**

Biography unknown.

Collection

(1) One box of Mss: 22 sheets of pencil drawings, one coloured, of pond life from sites in west London, 1886–1889.
L MSS BOL

18. **BRIDSON, Gavin Douglas Ruthven (b.1936)**

Bridson worked for Quaritch, the antiquarian bookseller, until joining The Natural History Museum as Zoology Librarian in 1963. In 1969 he moved to the Linnean Society (q.v.) as Librarian and Archivist, before taking up the post of Bibliographer and Senior Research Scholar at the Hunt Institute for Botanical Documentation in 1982. Bridson is author, with Geoffrey Wakeman, of *Printmaking and Picture Printing: a Bibliographical Guide* (1984).

Biographical Reference: personal information.

Collection

(1) Two volumes of Ts and a card index: *Rough Compilation of Materials Towards a Historical Directory and Bibliography of 19th Century Local Natural History Societies and Institutions,* [c.1976]. This collection is on loan from the author.
L MSS BRI

19. BRITISH ASSOCIATION FOR THE ADVANCEMENT OF SCIENCE

Founded in 1831 'to make the cultivators of science acquainted with each other, to stimulate one another to new exertions, to bring the objects of science more before the public eye and to take measures for advancing its interests and accelerating its progress'. The Association holds a meeting each year in one of the provincial cities of the British Isles.

Historical Reference: Howarth, O. J. R., 1922. *The British Association for the Advancement of Science: a Retrospect 1831–1921*. 318pp. London: The Association.

Collection

(1) One Ms volume: *Minute Book of the British Association Committee for the Compilation of an Index Animalium,* 1896–1912. The volume contains inserted letters, printed cuttings and notes. Mss in a wooden cabinet: the original paper slips prepared by C. D. Sherborn, from which *Index Animalium* was printed, together with additional unpublished notes, [c.1890–1935]. Presented by the British Association, 1912.
L MSS BRIT

20. BROADLEY, Alexander Meyrick (1847–1916)

Broadley was a lawyer who worked for many years in Tunisia and Egypt, and who wrote numerous historical novels and romances. He was a collector of books, prints, coins and medals, whose collections were auctioned in 1918.

Biographical Reference: 1891. *Supplement to Allibone's Critical Dictionary of English Literature*. 2 vols. Philadelphia.

Collection

(1) One printed volume: *The Life of Sir Joseph Banks* by Edward Smith (1911), extra-illustrated with four letters to or from Banks, 1773–1802, and one letter of Sir James E. Smith. Five letters from the author to Broadley, 1910–1911, are also tipped into the book. Purchased 1918.
L MSS BROA

21. **BROOKE, Winifred May A. (1893–1975)**

Miss Brooke was born in Stroud Green, Hampshire. She was a trained artist with an interest in natural history, particularly botany. She travelled widely, and collected plants in many parts of Europe, South Africa, South America and the East Indies. She published *Sketches from Nature in the North of Ireland* (1935), and presented herbarium specimens to the Museum.

Biographical Reference: 1976. Obituary. *Biological Journal of the Linnean Society* 8:368.

Collections

(1) Thirteen boxes of drawings in pencil and pen and ink, many coloured. In four series: botanical, 669 sheets, including South Africa (1936–1938), Switzerland (1947), fungi (undated); zoological, 595 sheets, including marine invertebrates, birds, domestic and farm animals (undated); entomological, 127 sheets (undated and unlocalized); general, 82 sheets, including scenery and buildings (undated). Presented by Mr J. Squier, 1980.
L MSS BRO

(2) Four boxes: 225 lantern slides of plants, animals, people and scenery in South Africa, Bolivia and elsewhere, [c.1925–1940]. Presented by Mr J. Squier, 1980.
L MSS BRO

22. **BUCKLAND, William (1784–1856)**

Buckland was Reader in Mineralogy (1813) and Geology (1818) at Oxford University. He was particularly interested in caves and cave faunas which, early in his career, he saw as evidence of a universal deluge. He was President of the Geological Society from 1824 to 1826 and again from 1839 to 1841. Buckland took holy orders in 1809, became Canon of Christ Church, Oxford, in 1825, and Dean of Westminster in 1845.

Biographical Reference: DSB.

Collection

(1) One Ms in a folder: letter to the Revd J. S. Henslow, August 1834, about the impending visit of Monsieur Arago. Purchased.
L MSS BUC

23. **CARPENTER, William Benjamin (1813–1885)**

Carpenter was born in Exeter and was educated in Bristol, before attending University College, London, and Edinburgh University as a medical student. He never went into general practice, but moved to London, where he lectured in physiology and forensic medicine. He became Registrar of the University of London in 1856, and devoted the rest of his working life to the development of that institution. Carpenter was a wide-ranging naturalist, author of many books and papers on zoological and palaeontological topics.

Biographical Reference: DNB.

Collection

(1) Two Mss in a folder: one letter to the Revd J. Crompton, 13 April 1871; one letter to the Revd J. E. Hall, 15 July 1856. Purchased 1977.
L MSS CARP

24. CARTER, Harold Burnell (b.1910)

Carter graduated from the University of Sydney, and was a research scientist with CSIRO in Australia, and later with the Agricultural Research Council in Edinburgh. He became interested in the life of Sir Joseph Banks (q.v.) through his work on animal production. He is author of *Sir Joseph Banks, 1743–1820* (1988) and Director of the Banks Archive Project.

Biographical Reference: personal information.

Collection

(1) Photographs in a folder: three photographs of Joseph Banks's rooms in Christ Church, Oxford, taken by Carter, 1972.
L MSS CART

25. CHALMERS-HUNT, J. Michael (b.1920)

Chalmers-Hunt is the author of many scientific papers on the Lepidoptera, as well as important reference works such as *Local Lists of Lepidoptera* (1989). He was editor of the *Entomologist's Record and Journal of Variation* from 1973 to 1985, and was twice President of the British Entomological and Natural History Society.

Biographical Reference: Chalmers-Hunt, J. M., 1989. *Local Lists of Lepidoptera*. 247pp. Uffington. Dust jacket.

Collection

(1) One box of Mss: eleven notebooks, notes, Tss and correspondence on natural history auctions, 1970–1976. The results were published in *Natural History Auctions, 1700–1972* (1976). Presented by J. M. Chalmers-Hunt, 1980.
L MSS CHA

26. **CHEESMAN, Lucy Evelyn (1881–1969)**

Cheesman was well known as a broadcaster and writer, as well as being an entomologist and traveller. She made many expeditions in New Guinea, the New Hebrides and other islands in the Pacific. Cheesman wrote many scientific papers and books about her travels. She published an autobiography, *Things Worth While*, in 1957.

Biographical Reference: 1969. Obituary. *Entomologist's Monthly Magazine* 105:217–219. Portrait.

Collection

(1) One Ms volume: *Gazeteer of New Guinea*, compiled for the Naval Intelligence Division, 1941, and published by the Admiralty.
L MSS CHE

27. **CHINA**

Collection

(1) One volume: 39 watercolour drawings by native artists of Chinese plants, reptiles, fish and birds, undated, but probably early nineteenth century. Presented by F. A. Philbrick, 7 November 1894.
L MSS CHI

28. **CLIFT, William (1775–1849)**

Clift was born and grew up in Cornwall. He moved to London in 1792 to become the pupil and amanuensis of John Hunter (1728–1793), and took charge of Hunter's museum at his master's death. When the Company of Surgeons took over the museum in 1799, Clift became the first Conservator, a post he held for forty-three years. Richard Owen (q.v.) became Clift's assistant in 1827, and married Clift's daughter, Caroline, in 1835.

Biographical Reference: *DSB*.

Collections

(1) One Ms volume: *Memoranda Relating to the Sale of Old and Duplicate Specimens of Natural History and Anatomical Articles by the Trustees of the British Museum to the Royal College of Surgeons in London, in the Year 1809*, 24pp., 1836.
L MSS CLI

(2) One letter in a folder: letter from Baron Georges Cuvier to Clift, undated. Presented by C. D. Sherborn.
L MSS CLI

(3) Six Mss in a folder: letter from Pedro de Angelis to Clift on *Glyptodon* and *Mylodon*, with notes by R. Owen, 1841; *Copy of Mr Hunter's MSS Account of the American Siren of Linnaeus*, [c.1834]; *Account of the Discovery of the Remains of Several Skeletons of the Mastodon, i.e. (Megatherium) in the Province of Buenos Ayres in South America* by Woodbine Parish Esq., 1832, with copies of letters to the Trustees of the British Museum; *On Muscle Fibre* by B. C. Brodie, undated; *Report of the Dissections of Animals, Diseased Parts, Monstrosities &c Presented to or Purchased by the Royal College of Surgeons*, six leaves, [1831]; *Fossil Edents S. America*, letters on the fossil Edentata of South America by J. B. Pentland, W. Parish and W. Buckland, with other descriptions, 1835–1838; all transcribed by Clift, or his son, William Home Clift.
L MSS CLI

29. **COLLIER, Frank Simon** (1900–1964)

Collier grew up in Hertfordshire and went to Oxford University. He travelled to Nigeria in 1923, where he served as a Forest Officer, and later Chief Conservator. He developed an unequalled knowledge of the fauna of West Africa, and was an accomplished natural history artist. Collier strove unsuccessfully to establish wildlife laws in Nigeria. He was on the Council of the Fauna Preservation Society from 1953 to 1957.

Biographical Reference: 1964. Obituary. *Oryx* 7:211.

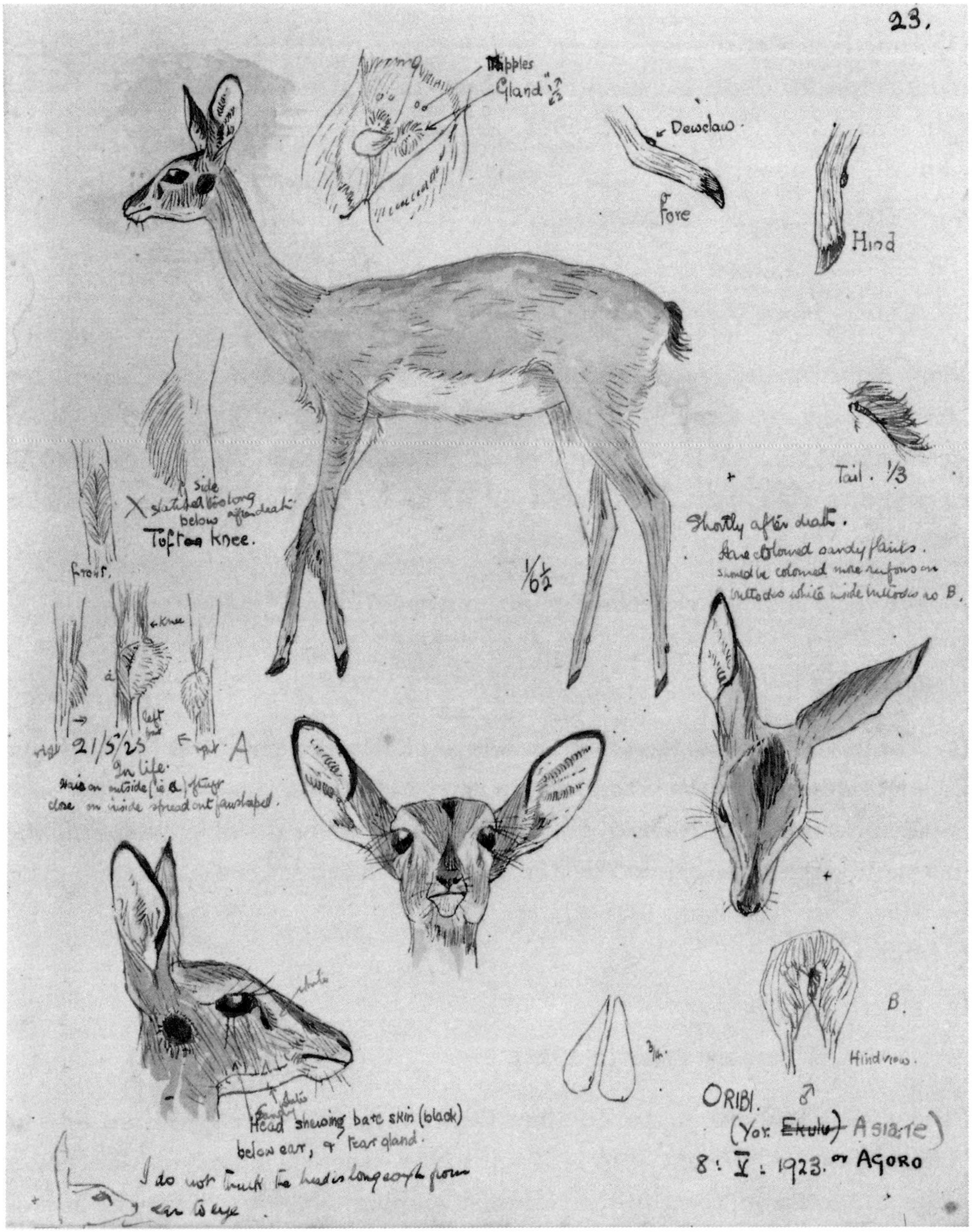

29(1) Collier, Frank Simon (1900–1964)
Sketches of an oribi illustrating diary entry for 8 May 1923, when Collier was in Olokemeji, Nigeria.

Collection

(1) Four boxes of diaries: eleven volumes and five folders of an illustrated natural history diary covering many parts of England, North Wales, British Cameroons and Nigeria, 1915–1935. A finding aid is available. Presented by Mrs M. A. Popay, 1982.
L MSS COL

30. CUST, Hon. Mary Anne (1799/1800–1882)

Mary Anne Boode grew up in Cheshire and married Edward, later General The Hon. Sir Edward, Cust (1794–1878) in 1821. He had seen active service in the Peninsular War, sat as a Member of Parliament and became Master of Ceremonies to Queen Victoria in 1845. Lady Cust wrote several books, including *The History and Diseases of the Cat* (1856).

Biographical Reference: *DNB* (for her husband).

Collection

(1) Three volumes: watercolour drawings of flowers, fruit, fish and marine invertebrates painted on a tour of the West Indies, January – July 1839, with detailed captions by E[dward] C[ust]; and paintings of flowers, fruit, fish and marine invertebrates painted in Tenerife and Madeira, 1866–1874. Presented by Miss Cust, February 1931.
L MSS CUS

31. DANCE, Stanley Peter (b.1932)

Peter Dance worked in the Zoology Department (Mollusca Section) of The Natural History Museum from 1957 until 1966, when he moved to Manchester Museum, and from there to the National Museum of Wales. He is the author of *A History of Shell Collecting* (1986) and *The Art of Natural History* (1978).

Biographical Reference: personal information.

Collection

(1) One letterbook: 81 letters from Warren Royal Dawson to Dance, 1962–1967. Presented by S. P. Dance 1985.
L MSS DAN

32. **DARWIN, Charles Robert (1809–1882)**

The most important event in Charles Darwin's life was his five-year voyage round the world aboard the Admiralty surveying ship HMS *Beagle*, under the command of Robert Fitzroy (q.v.). The observations and collections made on this voyage led Darwin to important publications on geology and botany. They also led to his interest in adaptation and the origin of species, from which he developed his theory of evolution by natural selection, published in his book, *On the Origin of Species* (1859).

Biographical Reference: *DSB*.

Collections

(1) One letterbook: 65 original and photocopied letters of Charles Darwin, 1843–1882. The 31 originals include letters to H. Anton de Barry (2, 1879), A. G. Butler (3, 1871–1879), Emma Darwin (1, 1844), W. E. Darwin (5, 1872), J. E. Gray (1, 1856), H. T. Stainton (9 and 3 replies, 1855–1881), G. R. Waterhouse (2, 1843–1860). A finding aid is available.
L MSS DAR A

(2) Ms in a folder: [On instinct], five leaves from the original manuscript of *On the Origin of Species*, [1859]. Presented by Mrs Litchfield, 1916.
L MSS DAR A

(3) Mss in a folder: descriptions of corals collected on the voyage of the *Beagle*, possibly written out by Darwin to accompany his paper on coral reefs, read to the Geological Society on 31 May 1837, [?1837]. Transferred from Coelenterata Section, 1975.
L MSS DAR A

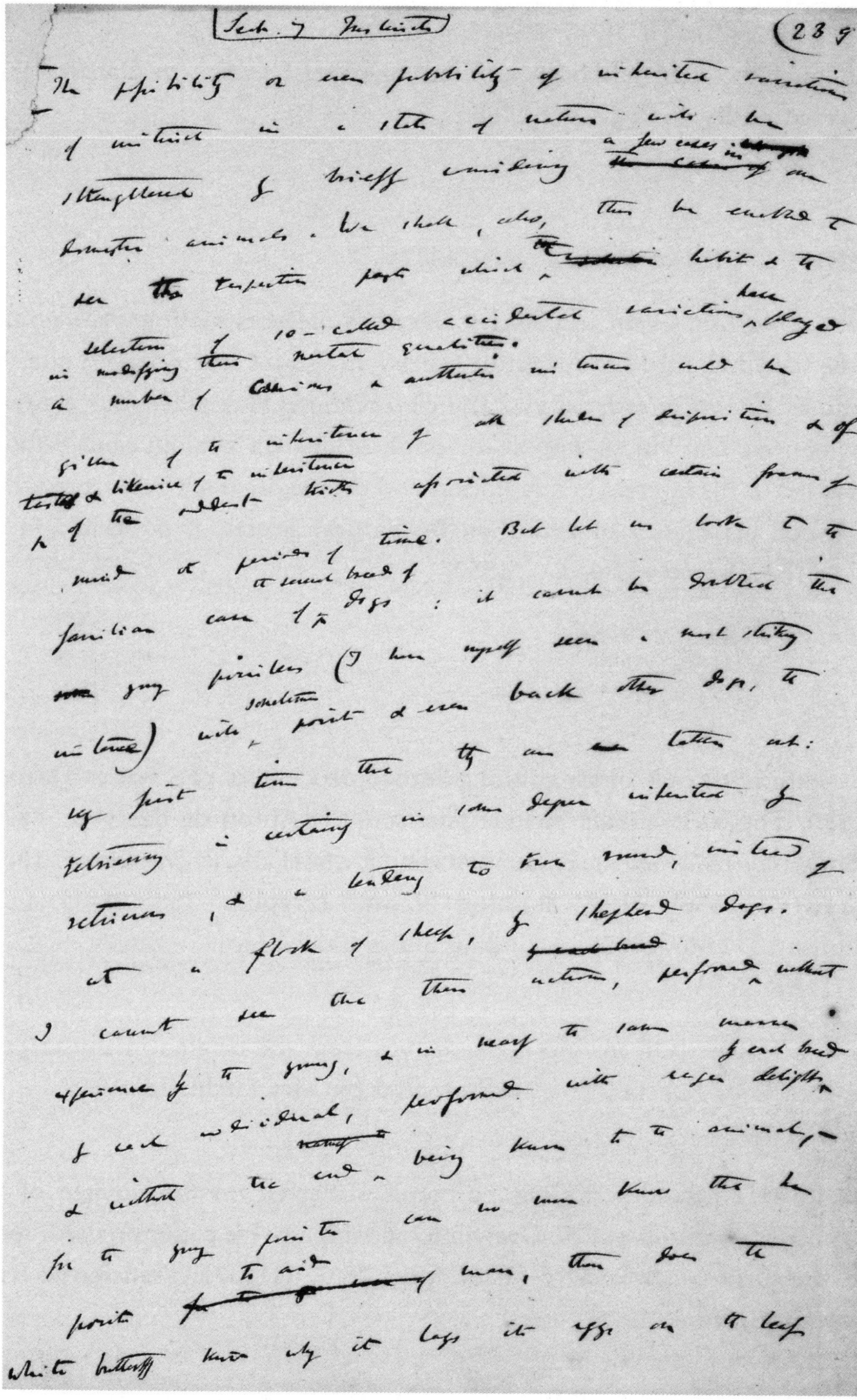

32(2) Darwin, Charles Robert (1809–1882)
One leaf of the original manuscript of *On the Origin of Species* (1859). The first sentence reads: 'The possibility or even probability of inherited variations of instinct in a state of nature will be strengthened by briefly considering a few cases in our domestic animals.'

33. DAVIS, Derek S.

On the staff of the Central Electricity Research Laboratories, Bradwell-on-Sea, Essex.

Biography unknown.

Collection

(1) Mss in a folder: lists and correspondence relating to natural history trips on the Blackwater Estuary aboard the research vessel *Cluaran*, organized in connection with the Council for Nature's youth and natural history theme, 1964.
L MSS DAV

34. DAVIS, Thomas Arthur Warren (1899–1980)

Davis was born and brought up in Pembrokeshire. He served in India during the First World War, and graduated in forestry from Oxford University in 1921. He worked as an Assistant Conservator of Forests for the Forestry Department in British Guiana before returning to India for Second World War service on the North-West Frontier. Davis retired from the Indian Forestry Service about 1948, and spent his long retirement in Wales, where he became an authority on the natural history of the Principality.

Biographical Reference: Obituary. *Nature in Wales* 17:157–158.

Collection

(1) Three boxes of Mss: eleven diaries, 1918–1943; general and *Strychnos* correspondence, 1932–1966; ornithological notes and observations in Iraq, British Guiana, India, Kashmir and Wales, 1919–1957; notes on the trees, vegetation, mammals and migration of butterflies in British Guiana, 1929–1950; photographs of British Guiana, 1925–1939. A finding aid is available. Bequeathed by the author, 1980.
L MSS DAVIS

35. **DAWSON, Muriel Helen (1897–1974)**

Muriel Dawson spent her early years in New Zealand, before moving to England in 1913, where she went to school in Richmond and later attended the Royal College of Art from 1918 to 1922. She worked on book illustrations for many years, and lived in the Shetland Islands from c.1959 until her death in 1974.

Biographical Reference: personal information.

Collection

(1) Five boxes and one folder of drawings: *Plant Studies*, watercolour drawings in a folder, 1918–1921; one sketchbook of watercolour drawings of butterflies and flowers, 1920; one sketchbook of watercolour views, butterflies, flowers and people, 1938–1969; eight sketchbooks and 22 folders and packets of pencil and watercolour sketches of birds, mammals and boats on and around the Shetland Islands, 1959–1974. A finding aid is available. Bequeathed by the artist.
L MSS DAW

36. **DAWSON, Warren Royal (1888–1968)**

Born in Ealing, London, and educated at St Pauls School. Dawson was Honorary Librarian to the Corporation of Lloyds from 1927 to 1948. He wrote papers and books on archaeology, Egyptology, the history of medicine and science, and prepared catalogues of a number of important manuscript collections, including the Huxley papers for Imperial College and the J. E. Smith papers for the Linnean Society (q.v.). Warren Dawson edited *The Banks Letters* (1958) for the Museum.

Biographical Reference: 1972. *Who Was Who, 1961–1970*. 1243pp. London.

Collections

(1) Nine boxes of letters: the correspondence of Warren Dawson on scientific, medical and historical topics, including his work on the correspondences of

Sir Joseph Banks, T. H. Huxley, and Sir James E. Smith; 2150 letters from more than 200 correspondents, including W. R. Bett (32, 1945–1967), R. Broom (34, 1945–1950), F. L. R. Brown (21, 1941–1954), H. B. Carter (106, 1955–1966), W. H. Curtis (39, 1942–1957), S. P. Dance (36, 1961–1967), G. R. De Beer (167, 1947–1961), J. A. Ewan (23, 1956–1966), L. F. Gilbert (91, 1953–1955), B. Henrey (28, 1954–1965), P. Hepworth (40, 1954–1966), J. W. F. Hill (75, 1950–1965), P. M. Jones (31, 1946–1962), L. G. Langwill (70, 1947–1966), W. R. Lefanu (73, 1945–1967), C. G. Lowry (30, 1945–1946), H. M. McKay (20, 1945–1950), K. P. Oakley (25, 1946–1965), S. Savage (49, 1941–1965), R. A. Skelton (39, 1954–1961), B. M. Skramovsky (102, 1954–1959), P. G. Summers (34, 1952–1955), A. C. Townsend (95, 1945–1963) and E. A. Underwood (28, 1946–1952). A finding aid is available. Presented by the British Library, 1978.
L MSS DAWS A

(2) Three boxes of Ms: *Calendar of the Letters of Sir Joseph Banks*, [c.1957], published as *The Banks Letters*, The Natural History Museum, 1958.
L MSS DAWS B

(3) One Ms volume: *Banks Correspondence Register of Names*, 1955.
L MSS DAWS B

(4) One volume of Mss: *Letters Relating to the Frazer Lectures*, 78 folios, 1932, including L. Frazer (26, 1932). 'Dawson Mss 16' This correspondence relates to his editorship of *The Frazer Lectures, 1922–1932* (1932).
L MSS DAWS C

(5) Three Mss in a folder: *Index to the Engraved Portraits in John Nichols 'Literary Anecdotes of the Eighteenth Century'*, five leaves, undated; *Index to the Portraits of Naturalists Published in John William Taylor's 'Monograph of the Land and Freshwater Mollusca of the British Isles'*, five leaves, undated. *Alphabetical List of the Lord Arthur Russell Collection of Portraits of Naturalists in the Library of the Linnean Society*, undated. Presented by Dawson, 1958.
L MSS DAWS D

37. DE BEER, Sir Gavin Rylands (1899–1972)

Born in Surrey, De Beer was educated in Paris, Harrow and at Oxford, where he graduated in zoology in 1921. He taught in the department until 1938, when he moved to University College, London, where he became Professor of Embryology in 1945. De Beer was Director of the The Natural History Museum from 1950 to 1960. He lived in Switzerland for much of his retirement. De Beer had an extraordinary range of expertise, from skull anatomy, embryology and evolution, to military history, early travellers in Switzerland and the work of Charles Darwin (q.v.). One of his most important books was *The Development of the Vertebrate Skull* (1937).

Biographical Reference: DNB.

Collection

(1) One Ts volume: *Atlas of Evolution*, 373pp. [c.1962]. Published by Thomas Nelson and sons, 1964. Transferred from the Department of Palaeontology, 1978.
L MSS DEB

38. DESHAYES, Gerard Paul (1795–1875)

Deshayes studied medicine in Strasburg and Paris, but never practised as a doctor. He settled in Paris as an independent researcher in natural history, particularly in recent and fossil conchology. He built up a huge collection of shells, which he eventually sold to the French government, and published *Description des Coquilles Fossiles des Environs de Paris* (1824–1837). In 1869 Deshayes became Professor of Conchology at the Muséum National d'Histoire Naturelle in Paris.

Biographical Reference: DSB.

Collection

(1) One bound Ms in a folder: *Conchiferes – Conchifera*, [c.1834], in French. Published in English in *The Cyclopaedia of Anatomy and Physiology*, edited

by R. B. Todd, 1836. Presented by C. D. Sherborn, 1908, formerly in the collection of Sir Richard Owen.
L MSS DES

39. DE WINTON, William Edward (1856–1922)

De Winton combined his business in London with a keen interest in natural history. He studied both birds and mammals, and contributed to the literature on both groups. His most important publication was the volume on Mammalia in *Zoology of Egypt* by John Anderson (1902). De Winton was briefly Superintendent of the Zoological Gardens in London.

Biographical Reference: 1923. Obituary. *Ibis,* 11th series, 5:162–163.

Collection

(1) Three volumes of photographs: 435 prints, taken on a hunting expedition from Berbera to Mombasa, passing through British Somaliland, Abyssinia, Kenya and Uganda, showing scenery, people and wildlife, 1896–1898. Presented by Mrs De Winton, 1968.
L MSS DEW

40. 'DISCOVERY'

A steam barque, built in 1901 for the National Antarctic Expedition. She worked in Antarctic waters until 1931, when she was laid up, before taking a berth on Victoria Embankment from 1937. *Discovery* is now open to the public in Dundee, the port where she was built.

Historical Reference: Rice, A. L. 1986. *British Oceanographic Vessels, 1800–1950*. 193pp. London: Ray Society, ref. 47–54.

Collections

(1) One wooden box of Mss: 15 registers, logbooks, diary and notebooks from the National Antarctic Expedition, 1901–1904: *Stations Log,* the record of 15 dredging operations; *Sea Temperature Register,* a record of observations by

M. Barne; *Collections Register I*, details of the botanical, freshwater and land collections, c.100 items; *Collections Register II*, marine collections, kept by Reginald Koettlitz, c.500 items; *Collections Register III* and *IV*, plankton samples, kept by Koettlitz, 288 collections; *Measurements of Crew*, with physiological notes, by Koettlitz; *Koettlitz Bacteria*, records of 64 samples, with four art originals; *Koettlitz Plankton*, tabulated results, rough notes and sketches; *Koettlitz Summary*, weekly summaries of his scientific work; *Hodgson Diary I* and *II*, a daily diary kept by Thomas Vere Hodgson, together with natural history and other notes; *Seal Stomachs*, notes by Hodgson; *Dredge & Trawl*, notes by Hodgson; *Memoranda for Discovery Sledge Party, Nov & Dec 1903*, notes by Hodgson.
L MSS DIS A

(2) One folder of photographs: 59 prints and two negatives taken by E. R. Gunther, A. C. Hardy and S. W. Kemp on the expedition to South Georgia to investigate the whaling industry, 1926–1927.
L MSS DIS B

41. **DITMAS, Frederick (1811–1876)**

Served in India in the Royal (Madras) Engineers from 1826 to 1850, being gazetted a Second Lieutenant in 1829. He was in Secunderabad in 1846 and 1847, and in Bellary in 1849. Ditmas was Professor of Fortification at the Honourable East India Company's Military College, Addiscombe from 1850 to 1861.

Biographical Reference: letter from E. M. R. Ditmas.

Collection

(1) One volume of 59 watercolour drawings of birds, insects and reptiles from India, captioned in pencil, [c.184–], bound with eight printed cards. Presented by Miss E. M. R. Ditmas, 1967.
L MSS DIT

42. FISCHER, Jacob Benjamin (1730–1793)

Author of a natural history of Livonia entitled *Versuch einer Naturgeschichte von Livland* (1778, 2nd edn. 1784).

Biography unknown.

Collection

(1) One Ms volume: *Zusatze zur Zwoten Auflage seiner Naturgeschichte von Livland*, 'additions to the second edition of his Natural History of Livonia', Riga, 28 February 1793. Fischer may or may not be the author of the additions. Presented by Dr A. Apinis, 1989.
L MSS FIS

43. FITZROY, Robert (1805–1865)

Fitzroy was born in London, a member of an aristocratic family. He attended the Royal Naval College at Portsmouth in 1819, and was commissioned as a Lieutenant in 1824. He commanded HMS *Beagle* on her two surveying expeditions, accompanied by Charles Darwin (q.v.) for the 1831–1836 voyage. On his return he was successively Member of Parliament for Durham (1841), Governor of New Zealand (1843) and Commander of the Woolwich Arsenal (1848). When he retired from active service he devoted himself to meteorology.

Biographical Reference: *DSB*.

Collection

(1) Ms in a folder: *Adml Fitzroy's Remarks on Dr Stark's 'Currents of the Atlantic'. 1859. and the Author's Reply*. The 'remarks' are accompanied by copies of letters of James Stark and Robert Russell, March 1859. Purchased 1970, and formerly in the collection of Sir Richard Owen.
L MSS FIT

44. FORBES, Edward (1815–1854)

Forbes grew up on the Isle of Man, where he became a keen naturalist. After attending medical school in Edinburgh, he joined Captain Graves as naturalist to a naval expedition in the Mediterranean, and carried out an intensive dredging programme. On his return to London, Forbes worked at King's College and the Geological Society, before being appointed Naturalist to the Geological Survey. He returned to Edinburgh as Professor of Natural History in 1854, and died shortly after. Forbes's most important writings were concerned with biogeography and its relation to geological changes.

Biographical Reference: *DSB*.

Collections

(1) Three boxes of notebooks: 18 field notebooks containing pencil notes and sketches on the geology and fossils of Dorset and the Isle of Wight, North Wales, the Welsh Borders and northern England; humorous sketches and caricatures are also included, [c.1845–1853]. A finding aid is available. Presented by Major R. Godwin-Austen, 1924.
L MSS FORB A

(2) One Ms in a folder: One letter from Forbes to James Scott Bowerbank, undated. Purchased 1977.
L MSS FORB B

45. FOX, Dr Charles Arundel Overbury

Dr Fox lived at Valley House, Bishopston, Swansea.

Biography unknown.

Collection

(1) One folder of Mss and Tss: transcripts of accounts of the death of Captain James Cook from the journal of David Samwell, the *United Services Magazine*, Zimmerman's *Account of the Third Voyage* and other sources, c.1950–1955. Purchased 1973.
L MSS FOX

46. GRANT, James

A surgeon, of Calcutta. Possibly the James William Grant (1801–1873) who became a Member of the Royal College of Surgeons in 1823.

Biography unknown.

Collection

(1) Ms in a folder: *Post mortem examination of a female orang outang, March 1829,* published in *Zoological Journal* 5:91–95, 1830. Purchased 1970, and formerly in the collection of Sir Richard Owen.
L MSS GRAN

47. GRAY, George Robert (1808–1872)

Born and educated in London, Gray was appointed an Assistant in the Zoology Branch of the British Museum in 1831. He worked on both insects and birds, publishing a number of official catalogues and other works. His most important book is *Genera of Birds* (1844–1849). He was the younger brother of J. E. Gray (q. v.).

Biographical Reference: *DNB*.

Collection

(1) One letterbook: 44 letters of naturalists and scientists, mostly addressed to G. R. Gray, 1833–1872, including a number of letters of testimonial in support of his application to succeed Charles Koenig as Head of the Mineralogical and Geological Branch of the British Museum. Purchased 1977.
L MSS GRAYG

48. GRAY, John Edward (1800–1875)

J. E. Gray was born in Walsall, and studied chemistry and medicine in London. He took up the study of botany as a young man, but turned to zoology later on. Gray worked with J. G. Children (1777–1852) at the British Museum from

1824, and succeeded him as Keeper of the Zoological Collections in 1840. He resigned in 1874, shortly before his death. As well as his numerous zoological publications, Gray wrote on postage stamps, coinage and other topics.

Biographical Reference: *DNB*

Collections

(1) One volume of Mss: *J. E. Gray Miscellaneous Papers*, a collection of holographs, transcripts and Tss gathered together by B. B. Woodward, and with additions by A. E. Gunther. Principal items include: autobiographical notes in the hand of E. A. Smith, with corrections by J. E. Gray, two versions, after 1867; criticisms of Francis Galton's *Hereditary Genius*, 1869, in the hand of E. A. Smith, with additions by J. E. Gray; autobiographical and family notes in the hands of E. A. Smith and E. J. Miers, undated; autobiography, first and second drafts, holograph, 1863; autobiographical fragment, 1873, in the hand of E. J. Miers; fair copy of Gray's application for the keepership, 1840, with testimonials; four letters from J. Winter Jones concerning Gray's resignation, 1874; letters from Gray to Baron Cuvier (1830), W. H. Flower (1869), Professor Owen (1857) and others.
L MSS GRAYJ A

(2) One Ms volume in folder: *Biographical Notes*, 1863, being the third draft of Gray's autobiography. Presented 1951.
L MSS GRAYJ B

(3) One Ms volume: *Autobiographical journal*, 1862, covering his family background and his life up to 1820, published by A. E. Gunther in *Journal of the Society for the Bibliography of Natural History* 7:35–76, 1974; autobiographical fragments, [c.1862–1868], covering episodes of his life during the years 1821–1861. Presented by Professor Peter Gray, 1970.
L MSS GRAYJ C

(4) One volume of pamphlets: *Library Commission 1850*, containing Gray's copies of printed papers on scientific institutions, the British Museum, its library and catalogue, 1819–1850. One item, *A Letter to the Earl of Ellesmere*

. . ., by J. E. Gray, 1849, contains his extensive Ms annotations. Presented by A. E. Gunther, 1972.
L MSS GRAYJ D

(5) Two Mss in a folder: one letter from Gray to Baron Georges Cuvier, 19 July 1829; one draft letter from Gray to T. H. Huxley, 30 Sept 1857.
L MSS GRAYJ E

49. GRAY, Samuel Frederick (1766–1828)

Samuel Frederick Gray was a pharmaceutical chemist in Walsall until about 1807, when he moved to London. He published an important pharmacopeia, as well as a number of botanical works, in which he was assisted by his son, J. E. Gray (q. v.).

Biographical Reference: DNB.

Collection

(1) One Ms volume: *Lectures on Plants and Insects,* probably delivered by Gray at Grainger's School of Medicine, Maze Pond, London, 1819–1820.
L MSS GRAYS

50. THE GUNTHER COLLECTION

The Gunther Collection relates to three generations of a distinguished zoological family. The bulk of the collection was purchased from Mr A. E. Gunther in 1969, and additional items were presented between 1973 and 1983. His arrangement is largely preserved in the numbering system given in this catalogue.
L MSS GUNTHER COLL

GUNTHER, Albert Carl Ludwig Gotthilf (1830–1914)

Albert Gunther was born and educated in Germany, where he developed an interest in natural history while studying theology and philosophy at

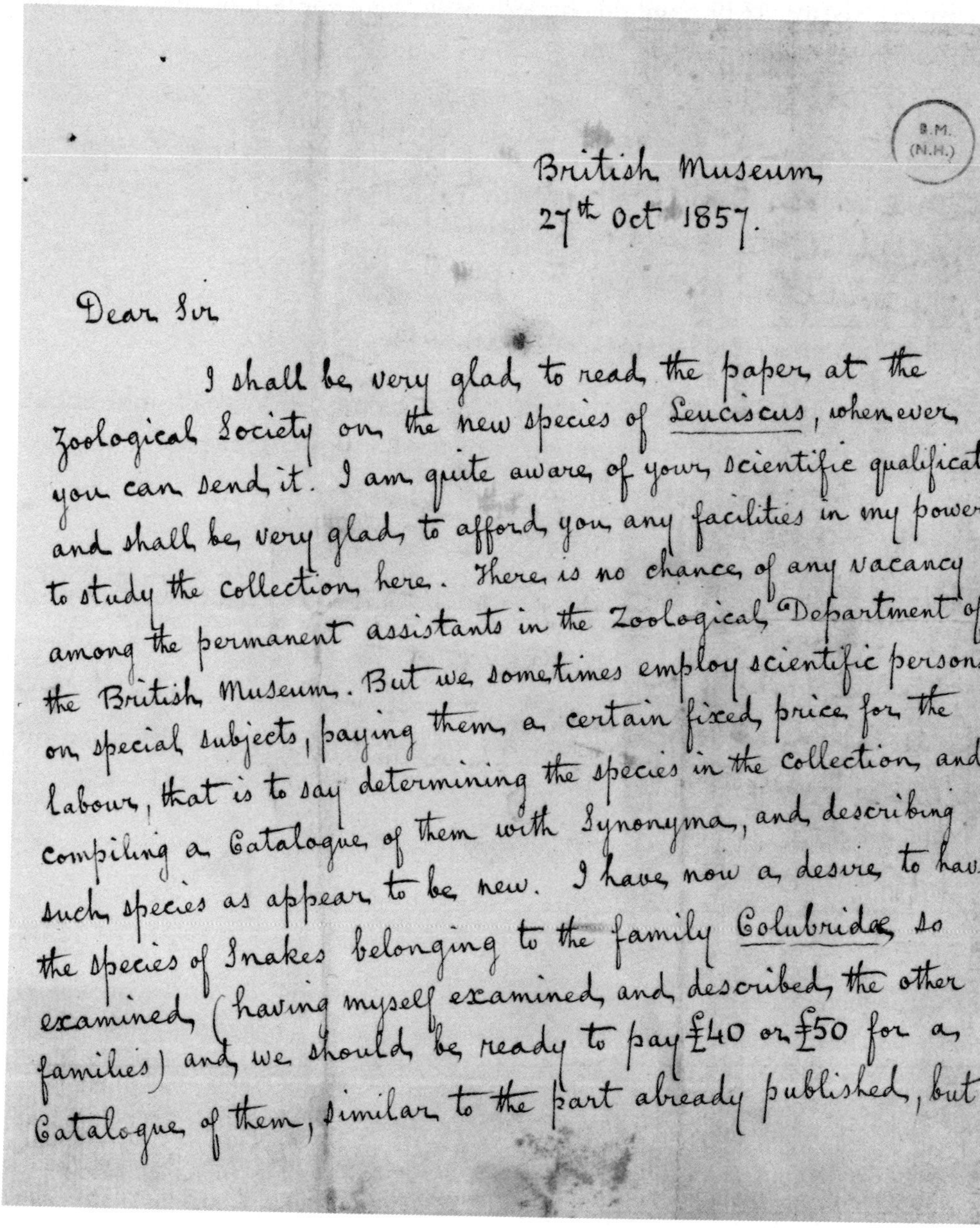

British Museum
27th Oct 1857.

Dear Sir

I shall be very glad to read the paper at the
Zoological Society on the new species of Leuciscus, whenever
you can send it. I am quite aware of your scientific qualification
and shall be very glad to afford you any facilities in my power
to study the collection here. There is no chance of any vacancy
among the permanent assistants in the Zoological Department of
the British Museum. But we sometimes employ scientific persons
on special subjects, paying them a certain fixed price for the
labour, that is to say determining the species in the collection, and
compiling a Catalogue of them with Synonyma, and describing
such species as appear to be new. I have now a desire to have
the species of Snakes belonging to the family Colubridae so
examined (having myself examined and described the other
families) and we should be ready to pay £40 or £50 for a
Catalogue of them, similar to the part already published, but

50(44) Gunther Collection
A letter from J. E. Gray, Keeper of Zoology at the British Museum, to Albert Gunther, offering him temporary work cataloguing the collection of snakes. Gunther accepted the job, and was taken on to the permanent staff in 1862. The letter was written by Adam White, an Assistant in the Zoology Department, and signed by Gray.

Tübingen. He moved to England in 1857, and was given a job in the Zoology Department of the British Museum, cataloguing the snakes. He succeeded J. E. Gray (q. v.) as keeper of the department in 1875 and served until his retirement in 1895. His greatest work was the *Catalogue of the Fish in the British Museum* (eight volumes, 1859–1870).

Biographical Reference: *DNB*.

GUNTHER, Robert William Theodore (1869–1940)

Born in Surbiton, the son of Albert and Louisa Gunther, Robert was educated in London and at Oxford University, where he graduated in zoology. He studied marine biology at Naples in 1893, before returning to Oxford as lecturer, and later tutor, in natural science at Magdalen College. Gunther developed an interest in the history of science and of scientific instruments, and wrote a number of volumes for the 'Early Science in Oxford' series.

Biographical Reference: *DNB*.

GUNTHER, Albert Everard (b.1903)

Graduating in geology at Oxford University in 1925, Gunther joined the Shell Oil Company and worked in California, Trinidad, Holland and Romania. He was involved in the restructuring of Germany's Geological Survey from 1945 to 1947, before returning to his career with Shell. After his retirement Gunther turned to the history of science, studying the work of his father and grandfather, and publishing *A Century of Zoology through the Lives of Two Keepers* in 1975.

Biographical Reference: Pfeiffer, D. 1987. Das Reichsamt für Bodenforschung in Celle 1945–1947. Geschichte-Arbeiten-Personen. *Geologisches Jahrbuch* A–102:3–11.

Collections

(1) Eight volumes: miscellaneous papers, 1853–1903, offprints, annotated proofs, newscuttings and pamphlets by Albert Gunther, with a Ms contents

list by R. W. T. Gunther in each volume. A few letters and Mss are also included.

(2) Six boxes: letters to Albert Gunther and R. W. T. Gunther, mainly scientific, 1865–1924, including L. Dollo (12, 1886–1908), E. Carnarvon (10, 1883–1896), G. King (10, 1900–1914), A. Morgan (17, 1861–1874), G. Neumayr (17, 1873–1906), J. W. Towse (11, 1883–1896).

(3) Four volumes in a box: diaries and a notebook, 1862–1903, comprising Albert Gunther's four official diaries for 1862–1874, and two books of memoranda for 1881–1883 and 1903–1905.

(4) One box: letters to Albert Gunther from British correspondents, 1860–1911, including G. D. Campbell, Duke of Argyll (14, 1869–1889).

(5) One box: letters to Albert Gunther from British naturalists, 1858–1911, including E. Tennent (14, 1861–1862).

(6) One box: letters to Albert Gunther from British zoologists, 1868–1912.

(7) Two boxes: letters to Albert Gunther from British Museum officers, 1869–1914, including F. J. Bell (13, 1896–1914), G. A. Boulenger (14, 1884–1914), A. G. Butler (20, 1895–1912), R. I. Pocock (10, 1892–1911), C. T. Regan (13, 1907–1913).

(8) One box: letters to Albert Gunther from early friends and relations, 1855–1912, including G. Benguerel (19, 1855–1856), G. Jaeger (26, 1856–1863), W. Widenmann (24, 1857–1868).

(9) Three boxes: letters to Albert Gunther and J. E. Gray from Continental correspondents, 1830–1913, including A. Koelliker (19, 1867–1904), R. Ledenfeld (10, 1886–1895), R. A. Philippi (10, 1895–1901), M. Weber (23, 1890–1913).

(10) One box: letters to Albert Gunther and J. E. Gray from foreign correspondents, 1831–1905, including M. H. Peracca (23, 1891–1898).

(11) One box: letters to Albert and R. W. T. Gunther and J. E. Gray from American correspondents, 1836–1930, including E. D. Cope (10, 1863–1888), T. N. Gill (14, 1861–1873), D. S. Jordan (14, 1884–1913).

(12) One box: letters to Albert Gunther from African travellers, 1868–1911, including R. Crawshay (14, 1895–1911).

(13) One box: letters to Albert Gunther from overseas travellers, 1859–1905.

(14) One box of Mss: salmon and trout correspondence, 1862–1908.

(15) One box: letters to Albert Gunther and others relating to Francis Day, 1865–1889.

(16) Twenty-one boxes and three volumes: letters to Albert and R. W. T. Gunther arranged alphabetically by correspondent, 1857–1930, including S. Alcock (32, 1890–1913), J. V. Barboza Du Bocage (24, 1863–1897), W. T. Blanford (13, 1875–1899), P. Bleeker (36, 1861–1873), Sir W. L. Buller (20, 1874–1889), W. W. Cole, Earl of Enniskillen (20, 1861–1886), J. Couch (26, 1860–1867), C. G. Danford (16, 1888–1908), T. De Grey, Baron Walsingham (28, 1878–1900), G. E. Dobson (15, 1875–1890), R. A. Dobson (11, 1895–1908), F. D. Drewitt (10, 1900–1913), H. J. Moreton, Earl Ducie (201, 1875–1914), Sir W. T. Thiselton Dyer (25, 1881–1908), Sir P. de M. G. Egerton (54, 1864–1880), C. A. Ellis (105, 1890–1906), Sir L. Fletcher (31, 1887–1914), Sir W. H. Flower (13, 1869–1898), W. Francis and son (25, 1867–1914), F. D. Godman (27, 1892–1914), Sir S. F. Harmer (23, 1908–1913), E. J. O. Hartert (34, 1892–1909), Sir J. D. Hooker (16, 1861–1909), G. B. Howes (10, 1873–1900), M. H. Kingsley (28, 1894–1899), J. L. G. Krefft (54, 1861–1874), St G. Littledale (13, 1891–1894), S. Long-Price (15, undated), R. J. Lowe (18, 1859–1864), E. C. von Martens (44, 1856–1904), St G. J. Mivart (12, 1873–1893), Sir John Murray (23, 1885–1911), A. Newton (153, 1859–1906), K. A. Oppel (11, 1857–1862), H. Owen (11, 1870–1874), Sir R. Owen (31, 1861–1880), D. Parker (102, 1891–1913), W. Peel (82, 1864–1883), W. C. H. Peters (55, 1859–1875), Sir R. L. Playfair (58, 1864–1897), T. H. Powell (27, 1875–1892), T. L. Powys, Lord Lilford (365, 1868–1896), W. L. von Rapp (11, 1857–1865), Lord W. Rothschild (172, 1898–1914), N. C. Rothschild (38, 1898–1913), O. Salvin (10, 1862–1897), P. L. Sclater (23, 1862–1898), J. T. Taylor (17, 1879–1903), F. H. Troschel (30, 1857–1875), R. B. N. Walker (26, 1898–1900), O. E. H. Wucherer (34, 1860–1872).

(17) One volume, printed: presidential addresses by Albert Gunther to the Linnean Society and the British Association, 1880–1900.

(18) One volume, printed: obituaries and bibliographies of Albert Gunther, 1913–1930.

(19) Two boxes: Mss, printed papers, letters and photographs relating to the giant tortoises of Aldabra, 1872–1903, including Lord Rothschild (52, 1897–1903).

(20) One box: letters, newscuttings and other printed papers from and relating to Sir E. Ray Lankester, 1884–1921, including Sir E. Ray Lankester (42, 1884–1908).

(21) One box: letters and printed papers relating to *Zoological Record,* 1865–1887, including C. S. Bate (22, 1865–1868), W. S. Dallas (53, 1865–1871), J. R. Greene (13, 1865), E. C. von Martens (42, 1865–1870), A. Newton (54, 1865–1871), E. C. Rye (12, 1870–1882), H. T. Stainton (11, 1870–1887) and E. P. Wright (14, 1865–1871).

(22) One box of Mss: birds and reptiles, including letters and papers on Museum catalogues and displays, and on the keeping of cage birds, 1882–1905.

(23) One box of Mss: miscellaneous papers and correspondence, 1875–1941, including New Teachers Bible correspondence, Royal Society correspondence on petitions regarding the British Museum, correspondence on fish and reptiles from Costa Rica, and correspondence on the disposal of parts of A. C. L. G. Gunther's library.

(24) Two boxes of Mss: correspondence relating to publications on fish, 1876–1910, including correspondence with C. Gerold on *The Study of Fishes,* correspondence with Macmillan & Co. on *A Guide to British Fishes,* and correspondence on *Fische der Sudsee, Journ Museum Godeffroy,* 1871–1910, including L. Friederichsen (133, 1873–1910), J. C. Godeffroy (26, 1871–1874) and J. D. E. Schmeltz (10, 1871–1876).

(25) One box: letters to Albert Gunther relating to the foundation of the Linnean Society's Hooker Medal, 1897–1898.

(26) One box of Mss: correspondence on the possible use of dried locusts as food for cage birds, 1891–1898.

(27) Two boxes of Mss: learned societies and national expeditions, Mss etc., 1875–1913, including printed ephemera and correspondence of the Linnean and Zoological societies, papers on the disputed election of the Secretary of the Zoological Society, 1901–1902, and papers on collections from the *Valorous, Challenger, Porcupine* and *Lightning* expeditions, 1875–1890.

(28) One box of Mss: *Encyclopaedia Britannica* correspondence, 1878–1902, including A. & C. Black (32, 1878–1887), G. H. Fowler (12, 1900–1902) and W. R. Smith (13, 1883–1887); *Biologia Centrali-Americana*, correspondence and proofs, 1890–1902.

(29) One box of Mss: the new Museum, Directorship and staff, Mss and memoranda, 1868–1884, including correspondence and papers on members of staff of the Zoology Department, on space and furniture requirements in the new building, on plans for zoological exhibitions, on the Index Museum, on reorganizing the Museum's administration, on Gunther's address to the British Association, 1880, and on Gunther's application for the post of Superintendent, 1884.

(30) One box of printed papers: Marine Biological Association, printed letters, reports and minutes relating to its foundation and early years, 1884–1894.

(31) One photograph album in a box: carte-de-visite portrait photographs, mounted in an album, with a contents list.

(32) One box of Mss: biographical notes of Albert Gunther by R. W. T. Gunther, with letters from students from Tübingen and a copy of his will, 1911.

(33) Six volumes and two boxes, printed: collected papers, reviews and monographs of Albert Gunther, 1866–1885.

(34) Three boxes of Mss: c.300 letters from Albert Gunther to R. W. T. Gunther, 1888–1914.

(35–39) Five volumes, printed: guides, regulations and other publications of the British Museum and The Natural History Museum, 1843–1890.

(40) One volume: Thames and fishery cases, comprising correspondence and

printed papers relating to Gunther's appearances as an expert witness, 1882–1898.

(41) One box of Mss: Metropolitan Sewages Discharge Commission Report, Ms, 1883.

(42) One box, printed: newspaper and magazine cuttings relating to the British Museum, The Natural History Museum and other topics, 1860–1936.

(43) One folder: photocopies of 41 letters from A. Gunther to Pieter Bleeker, 1861–1873 (Originals in Leiden Museum).

(44) One box and one folder: Ms and printed items selected to illustrate episodes in the life of Albert Gunther by A. E. Gunther, 1851–1914, including papers on his appointment, promotions and retirement from the British Museum; his *Reptiles of British India* and *The Study of Fishes*; naturalization papers, passport and insurance documents; completed questionnaire for Galton's *Inquiry into the Antecedents of Scientific Men*; first drafts for an autobiography; last will and testament and obituaries; and correspondence and papers relating to the death of Roberta Gunther, Albert's wife, in 1869, and his relations with his brother-in-law, W. C. McIntosh.

(45) One box and one folder: diplomas, certificates and testimonials, 1844–1913.

(46) One folder: museum plans, 1873–1882, including plans and perspective views of The Natural History Museum and museums in Edinburgh, Liverpool, Oxford, Vienna and elsewhere.

(47) One box: plans, elevation and detail of the spirit building, A. Waterhouse, 1882.

(48) Three Ms notebooks: notes of Professor Vierordt's lectures on physiology; notes of W. L. von Rapp's lectures on comparative anatomy; botanical notebook, 1848–1851.

(49) Twelve volumes in a box: Albert Gunther's account books, 1865–1913.

(50) Four volumes of Mss: notes for the life of A. Gunther by R. W. T. Gunther, with annotations by A. E. Gunther [c.1935–1940].

(51) Fifteen volumes, printed: *Catalogue of the Fishes in the British Museum* (1859–1870), interleaved and annotated by Albert Gunther, with letters, printed papers and other items inserted.

(52) Three volumes, printed: books from the library of Albert Gunther and his family, 1858–1878.

(53) One box: photographs and drawings of fish and marine mammals, with accompanying letters, 1886–1906.

(54) One box of Mss: *Fische der Sudsee* by Albert Gunther, Ms and proofs, 1909.

(55) One volume in a box: *Catalogue of Colubrine Snakes in the Collection of the British Museum*, interleaved and annotated by Albert Gunther, with Ms and printed insertions, 1858.

(56) Three letterbooks: correspondence of Louisa Eleanora Gunther, Albert Gunther's mother, and other members of the family, 1795–1860, 334 letters, including A. Gunther (122, 1847–1852), L. Mosthaf (27, 1827–1835), E. Nagel (48, 1830 and undated), K. Nagel (31, 1826–1835), F. Ostertag (38, 1823–1837).

(57) Two boxes, three volumes and a folder: biography of Albert Gunther by A. E. Gunther, notes, correspondence and Ts, 1965–1987.

(58) Two boxes and one folder of letters: R. W. T. Gunther correspondence, 1886–1925, comprising 167 letters received while a student at Oxford, 1886–1892, including L. E. Gunther (12, 1888–1891), H. Hamersley (10, 1889–1891), and E. Thompson (10, 1886–1892); letters and testimonials relating to Gunther's application for the keepership of Zoology, The Natural History Museum, 1908; letters relating to the death of Albert Gunther, 1914; and miscellaneous letters, including B. Tucker (12, 1921–1925).

(59) Three boxes of Mss: R. W. T. Gunther, scientific and historical Mss, 1888–1920, comprising notes for a book on the zoology of southern Italy, 1908–1909; scientific Mss, including notes on the Chaetognatha and siphonophores, a transcript of John Sibthorp's diary for his first trip to Greece, 1786–1787, notes on *Mnestra parasites*, on invertebrate zoology, and on animal distribution, reports on work at the Zoological Station in Naples,

1893–1900; notes on zoology and the history of zoology taken from books and lectures [1888–1920].

(60) Five volumes, printed: R. W. T. Gunther, annotated books and papers. *Lake Urmi, Persia*, offprints on the geography and natural history of Lake Urmi, with inserted letters, newscuttings, and sketches, 1888–1900; *The Phlegraean Fields*, offprints on the topography of the Phlegraean Fields, with inserted letters, cuttings, sketches and photographs, 1896–1924; *Naples, St Andrews*, printed papers on the Marine Zoological Station, Naples and the Gatty Marine Laboratory, St Andrews, with inserted letters and cuttings, 1890–1907; a guidebook to Persia, incomplete, annotated by Gunther before and during his travels in 1898; offprints of historical papers by Gunther, 1929–1930.

(61) Two boxes, a folder and five volumes: A. E. Gunther, papers relating to *Founders of Science at the British Museum* (1975) and *A Century of Zoology at the British Museum* (1980).

(62) Nine boxes and a folder: A. E. Gunther, notes and Mss on early Trustees and staff of the British Museum, 1968–1984, including papers on Thomas Birch (1705–1766), Matthew Maty (1718–1776), Edward Whitaker Gray (1748–1806), George Shaw (1751–1813), William Elford Leach (1790–1836) and John George Children (1777–1851).

(63) Five boxes, three volumes and three folders: A. E. Gunther, notes and Mss on John Edward Gray, 1968–1983.

(64) One volume: A. E. Gunther, summary of events in the life of R. W. T. Gunther, 1869–1940, prepared for *Early Science in Oxford*, volume 15, but not published, 1961–1964.

(66) Two boxes, eleven volumes and a folder: A. E. Gunther, notes and Mss on William Carmichael McIntosh (1838–1931), 1973–1977.

(67) Two packets: A. E. Gunther, autobiography (closed access).

(68) One volume, printed: A. E. Gunther, offprints of historical papers, 1970–1980.

(69) Three folders and four volumes: A. E. Gunther, papers, books and correspondence, 1937–1980, including letters from Newton K. Gregg about a reprint of Albert Gunther's *The Fishes of Zanzibar* (1967), notes on the education of naturalists and others, and two letters from Sir Hugh Elliott.

(70) Two volumes: Gunther Family Record, 1854–1914, comprising autobiography of Albert Gunther, transcribed by R. W. T. Gunther, 1914, and annotated by A. E. Gunther; synopsis of an autobiography of Albert Gunther, up to 1875, Ms and typed transcript; 'How I came to England' by Louise Gunther, typed transcript by R. W. T. Gunther; biographical notes on Carl Theodor Gunther by his brother Albert, 1911, Ms and typed transcript; Louisa Gunther's appointments book, 1854–1856.

(71) Four folders of prints and drawings for Albert Gunther's publications, including the *Report on the Deep-Sea Fishes Collected by HMS Challenger during the years 1873–1876* (1887) and *Fische der Sudsee*, undated.

51. **GURNEY, John Henry (1819–1890)**

Born in Norfolk, and educated at the Friends' School, Tottenham, Gurney worked in the family bank in Norwich for most of his life. He had a wide interest in natural history, and was an acknowledged expert on the birds of prey. His collection of these birds is housed in the Castle Museum, Norwich. Gurney published many articles and short notes in *Ibis*, which he helped financially for many years. He lived in retirement at Northrepps, near Cromer, Norfolk.

Biographical Reference: 1890. Obituary. *Ibis*, 6th series, 2:392–396.

Collection

(1) Letters in a folder: six letters from E. W. H. Holdsworth and six letters from W.. V. Legge, 1872–1878.
L MSS GUR

Quito 24 August 1853.

My dear Sir,

Below is a bad sketch, (although the colouring of the soft parts is correctly represented) of the species of Toucan brought to Europe by Bourcier. It is an inhabitant of the forest on the western declivity of the Andes – from 6000 to 10000 feet of elevation, and feeds on the fruits of several species of *Thibaudia*, many of which are of the size and shape of an olive – They generally go in pairs, male and female, and the plumage of both sexes is exactly similar – [struck out] – I have specimens of both

52 Handwriting Collection
Part of a letter from William Jameson (1796–1873), the Professor of Chemistry and Botany at the University of Quito, Ecuador, to the ornithologist John Gould (1804–1881). The letter concerns a number of birds that Jameson had collected to send to Gould.

52. HANDWRITING COLLECTION

This collection, containing examples of the handwriting of more than 3500 naturalists of the eighteen, nineteen and early twentieth centuries, was started in the General Library about 1900 as a result of prompting by Charles Davies Sherborn (q. v.). Sherborn presented many autograph letters from his own correspondence, and persuaded his friends C. W. Andrews, William Carruthers, T. R. Jones, Dr Henry Woodward and others to do the same. Older letters were added from the correspondence of a former Keeper of Geology, G. R. Waterhouse (q. v.). A large number of signed photographs were also incorporated into the collection by Sherborn. Letters are still occasionally added to the Handwriting Collection, but no attempt has been made to systematically collect autographs of late twentieth century naturalists. The collection is used from time to time to identify the authorship of unsigned manuscripts, specimen labels and incomplete letters, as well as for the considerable historical value of some of the letters.

53. HARMER, Frederic William (1835–1923)

Harmer was a businessman who devoted all his spare time to the study of the Crag deposits which make up the Pleistocene of East Anglia. He was the author of *Monograph of the Pliocene Mollusca* (1914–1923).

Biographical Reference: 1923. Obituary. *Geological Magazine* 60:285.

Collection

(1) One autograph album containing 86 signatures cut from letters addressed to Harmer and his geological collaborator, Searles V. Wood jun.
L MSS HARM

54. HARVEY-GIBSON, Robert John (1860–1929)

Harvey-Gibson was born in Aberdeen, and studied at the universities of Aberdeen, Edinburgh and Strasburg. He worked in the departments of Zoology and Physiology at Edinburgh, before moving to Liverpool, where he became

Lecturer in Botany. He was elected to the new Chair of Botany in 1894, and remained in post until 1921. Harvey-Gibson published a number of papers on algae, as well as *Outlines of the History of Botany* (1919).

Biographical Reference: 1929. Obituary. *Proceedings of the Royal Society of Edinburgh* 49:384–385.

Collection

(1) One volume, printed: *The Master-Thinkers*, 1928, a corrected and annotated proof copy. Purchased 1948.
L MSS HARVEY

55. HARVIE-BROWN, John Alexander (1844–1916)

Born in Stirlingshire and educated at Merchiston and the universities of Edinburgh and Cambridge. Harvie-Brown travelled in Norway and Russia in 1871–1872, and made several natural history cruises around the coasts of Scotland. He collaborated with T. E. Buckley (1846–1902) on surveys of the vertebrates of Scotland, and made a particular study of bird migration. He was the author of *Travels of a Naturalist in Northern Europe* (1905).

Biographical Reference: 1916. Obituary. *Ibis*, 10th series, 4:637–640.

Collection

(1) One folder of letters: five letters to Mrs Ratcliff, 1897–1901, on the papers of Sir William Jardine, her father.
L MSS HARVIE

56. HERSCHEL, Sir John Frederick William (1792–1871)

Herschel was educated at Eton and Cambridge University, where he read mathematics and passed out as a senior wrangler in 1813. He followed his father as an astronomer, although he also became interested in geology, crystallography, chemistry, meteorology and other scientific subjects. Herschel set up an observatory in South Africa from 1834 to 1838 and made pioneering observations of the southern skies. He was President and a medallist of the

Royal Society, and three times President of the Royal Astronomical Society. One of his best-known books is *Discourse on the Study of Natural Philosophy* (1830).

Biographical Reference: *DSB*.

Collection

(1) One Ms in folder: *Tick. Tick. Tick.*, 1865, a poem addressed to Lady Bell by Herschel, 'after her Grandfather's Clock'.
L MSS HER

57. **HINDS, Richard Brinsley (c.1812–c.1847)**

Hinds was surgeon-naturalist on HMS *Sulphur* from 1835 to 1842, and author of *Zoology of the Voyage of H.M.S. Sulphur* (1844). He died in Perth, Western Australia.

Biographical Reference: Desmond, R. 1994. *Dictionary of British and Irish Botanists and Horticulturalists*. 825pp. London.

Collection

(1) Two Ms volumes: diaries of the voyage of the *Sulphur* around the Pacific islands, December 1835 to July 1840. Purchased 1947.
L MSS HIN

58. **HOSE, Charles (1863–1929)**

Born in Hertfordshire, and educated at Felsted and Cambridge University, Hose joined the Sarawak civil service in 1884. He spent his whole career in the service, retiring and returning to England in 1907. He rose to be Divisional Resident and Judge of the Supreme Court. Hose was a keen naturalist, a field observer and collector, an ethnographer and cartographer. He was the first to spot the connection between the disease beriberi and diet. He published *Fifty Years of Romance and Research* in 1927.

Biographical Reference: *DNB*.

Collection

(1) Photograph album in a box: 170 black and white prints of photographs of the peoples, animals and plants of Sarawak, taken by Charles Hose and Robert Shelford, published by Valentine's Photographic Department, undated. Presented by Bedfordshire County Library, 1947.
L MSS HOS

59. **HUME, Allan Octavian (1829–1912)**

Hume was born on 6 June 1829, and educated at Haileybury and London University. He joined the Bengal civil service in 1849, and served in the North-West Province. He moved to the Revenue and Agriculture Department of the central government of India for a spell, before retiring in 1882. Hume returned to England in 1894. He was a keen naturalist, specializing in ornithology, and presented his important collection of Indian birds and eggs to the Museum in 1885. He published *Nests and Eggs of Indian Birds* (1873).

Biographical Reference: *DNB*.

Collections

(1) Six notebooks in a box: six natural history notebooks recording expeditions in different parts of India, 1867–1881, largely concerned with ornithology. Presented by W. E. Imeson, 1912.
L MSS HUM

(2) Two notebooks in a folder: two natural history notebooks recording further expeditions in India, 1881 and undated. Presented 1925.
L MSS HUM

(3) Printed papers in a folder: newscuttings on the game birds of India [by Hume] and on the Hume Collection in the British Museum, 1866–1891.
L MSS HUM

(4) Rolled document: illuminated addresses to Hume from the people of Berar and Surat expressing their gratitude and admiration, 1893 and undated.
L MSS HUM

60. **HUNT, Robert (1807–1887)**

Born in Plymouth, Hunt received training in science and medicine in London and the West Country. He worked for a chemical manufacturer in London and studied the chemistry of photography. He moved to Falmouth in 1840 and carried out researches on light and electricity. In 1845 he was appointed Keeper of Mining Records at the Government School of Mines in London, and spent the remainder of his career as Keeper and Professor at the school. Hunt was the author of *British Mining* (1884).

Biographical Reference: DNB.

Collection

(1) One volume, printed: a proof copy of Hunt's contributions to Lovell Reeve's *Men of Eminence in Literature, Science and Art* (1863–1867), bound with letters and Mss. A Finding aid is available. Purchased 1948.
L MSS HUN

61. **HUXLEY, Thomas Henry (1825–1895)**

Huxley was born in Ealing, West London, and was largely self-educated. He trained as a surgeon, and joined the Royal Navy, serving as Assistant Surgeon aboard the *Rattlesnake* on her voyage to the south-western Pacific, 1846–1850. He studied marine biology on his return, and published on the medusae. Huxley took up posts at the Royal School of Mines and the Geological Survey of Great Britain in 1854, and turned to vertebrate palaeontology and embryology. He is best known for his support for Charles Darwin (q. v.) in the controversy which followed publication of *On the Origin of Species* (1859).

Biographical Reference: DSB.

Collection

(1) One Ms in a folder: sketch of flower, insects, birds etc., drawn during a meeting of the Executive Committee of the City and Guilds of London Institute, 21 January 1884. Presented by Miss M. Parker, 1929.
L MSS HUX

62. INTERNATIONAL BIOLOGICAL PROGRAMME

The idea of an International Biological Programme was first discussed in 1959, as a result of the success of the International Geophysical Year. The formal inauguration of the programme took place at its First General Assembly in July 1964.

Historical Reference: Worthington, E. B. (ed.). 1975. *The Evolution of IBP*. 268pp. Cambridge: Cambridge University Press.

Collection

(1) One box of Mss: notes on the zooplankton and benthic faunas of Lake George, Uganda, by J. Darlington and others, 1968–1971. Presented by Dr P. H. Greenwood, 1975.
L MSS INT

63. JAPAN

Collection

1. One volume of drawings in a box: 200 watercolour drawings on silk, depicting insects, snails, worms, amphibians and reptiles, by a Japanese artist, [mid nineteenth century]. The 20 sheets of silk are mounted on stiff boards, and the volume has an embroidered binding. Presented by Sir Eric Miller, 1954.
L MSS JAP

64. JARDINE, Sir William (1800–1874)

Born in Edinburgh, Jardine was educated in York and at the University of Edinburgh, where he attended lectures in literature, medicine and natural history. Jardine succeeded his father as 7th Baronet in 1820. He was a prolific writer, and edited magazines and journals as well as *The Naturalist's Library* (1833–1845). He established a valuable museum at Jardine Hall.

Biographical Reference: *DNB*.

Collection

(1) One letterbook: correspondence for the year 1852, comprising 66 letters from 29 correspondents on a variety of scientific and social subjects. Presented by H. S. Gladstone, 1920.
L MSS JAR

65. **KINNEAR, Sir Norman Boyd (1882–1957)**

Born and educated in Edinburgh, Kinnear worked first as a volunteer at the Royal Scottish Museum. He was the Officer in Charge of the Bombay Natural History Society Museum from 1907 until 1920, when he joined The Natural History Museum as an ornithologist. Kinnear became Keeper of Zoology in 1945, and served as Director from 1947 to 1950. He was President of the British Ornithologists' Union, 1943–1948, and published papers in *Ibis*.

Biographical Reference: *DNB*.

Collections

(1) Two letterbooks: correspondence, comprising 58 letters from G. E. H. Barrett-Hamilton, two from F. G. D. Drewitt and 33 from J. A. Harvie-Brown, 1901–1923.
L MSS KIN A

(2) One Ms volume: *Journal of a Whaling Voyage in SY Scotia to the Greenland Seas, 15th April to 6th August 1907*, 1907. Presented 1947.
L MSS KIN B

(3) Folder of Mss: notes on the drawings of J. G. Loten (1710–1789), c.1925–1935. Presented by Lady Kinnear, 1957.
L MSS KIN C

66. KNIGHT, Thomas Andrew (1759–1838)

Knight was born at Downton Castle, Shropshire, and educated in Ludlow and at Oxford. He carried out experiments on breeding fruit and vegetables as a young man, and met Sir Joseph Banks (q. v.) in 1795. Knight published important papers on agriculture and horticulture in the *Philosophical Transactions of the Royal Society* between 1795 and 1806.

Biographical Reference: *DSB*.

Collection

(1) Letterbook in a box: 37 letters to Sir Joseph Banks, with a paper entitled *Descent of the Sap through the Bark. Gravitation one of the Causes,* 1796–1805. Purchased 1973.
L MSS KNI

67. LAISHLEY, Richard (1815–1897)

Laishley was born in Southampton and studied painting at the Royal Academy School, London. He emigrated to New Zealand in 1860, and became Minister to the Congregational Church at Onehunga. Apart from a short stay in Melbourne, Australia, he remained in New Zealand for the rest of his life. Laishley was an accomplished artist, and drew and painted insects, reptiles, birds, plants and landscapes.

Biographical Reference: Sibson, R. B., 1983. Richard Laishley, 1815–1897, priest, painter, naturalist. *Notornis. Ornithological Society of New Zealand* 30:29–33.

Collection

(1) One Ms volume: *Extracts from a Short Diary Written for a Grandchild,* 1890. Presented 1950.
L MSS LAI A

68. LINNEAN SOCIETY OF LONDON

Founded 1788, the Linnean Society is the oldest active biological society in the world.

Historical Reference: Gage, A. T. and Stearn, W. T., 1988. *A Bicentenary History of the Linnean Society of London*. 242pp. London Academic Press.

Collection

(1) One printed booklet in a box: *List of Fellows*, 1894, with Ms corrections and additions up to c.1925.
L MSS LIN

69. LOTEN, Joan Gideon (1710–1789)

Loten was the Dutch Governor of Ceylon from 1752–1757. He joined the Dutch East India Company in 1732 and sailed for Batavia. He rose up through the ranks, serving in Samarang, Macassar and Batavia before being appointed Governor of Ceylon. While working in the East he commissioned Pieter Cornelis de Bevere (b.1722), a land surveyor, to paint from life some of the mammals, birds, insects and plants in Ceylon and the Malay archipelago. Loten returned to Holland in 1758 and made the drawings available to several English naturalists and artists including Sydney Parkinson (?1745–1771), Peter Brown, George Edwards (1694–1773) and Thomas Pennant (1726–1798). He was made a Fellow of the Royal Society in 1760.

Biographical Reference: Ferguson, D. 1908, Joan Gideon Loten, FRS, the naturalist Governor of Ceylon (1752–1757) and the Ceylonese artist, de Bevere. *Journal of the Ceylon Branch of the Royal Asiatic Society* 19: 217–271.

Collection

(1) Two boxes: 154 watercolour drawings of mammals, birds, insects and plants painted by Pieter Cornelis de Bevere. Some of the drawings have Ms notes in Dutch by Loten. Purchased in 1925, having been bequeathed by Loten

to the Hollandsche Maatschappij der Wetenschappen, Haarlem, and later in the possession of Mr P. J. van Houten of Haarlem.
L MSS LOT

70. **McMILLAN, Nora Fisher (b.1908)**

Born in Belfast, Miss Nora Fisher worked in Belfast Municipal Museum (1929–1933), Liverpool Museum (1933–1937), and part time in the Geology Department of Liverpool University (1940–1956). She married in 1937 and became Mrs Nora Fisher McMillan. Mrs McMillan returned to Liverpool Museum in 1954, and continued to work there after her retirement. She is an authority on recent and Quaternary molluscs, as well as being an amateur botanist and naturalist.

Biographical Reference: personal information.

Collection

(1) One box: 350 letters from 250 conchologists, geologists, archaeologists and others, collected as examples of handwriting, 1853–1993. Most letters are addressed to Mrs McMillan, but others are to J. W. Jackson, J. Le R. Tomlin and R. Welch. Presented by Mrs McMillan, 1993–1994.
L MSS MCM

71. **MARTENS, Conrad (1801–1878)**

Martens was born in London, the son of a Hamburg merchant. For most of his life he worked as a professional landscape painter, acquiring some latterday fame from his association with Charles Darwin (q.v.) when he served with the *Beagle* from 1833 to 1834. Although the encounter was brief, the two remained in contact for many years. Martens left the *Beagle* at Valparaiso, and arrived in Sydney in 1835. He was entranced by Sydney harbour, and made his home in the city. Martens earned enough from sales of his landscapes to make a living for himself and his family.

Biographical Reference: J. Kerr (ed.) 1992. *Dictionary of Australian Artists.* 889pp. Melbourne.

Collection

(1) One box: seven pencil drawings of Australian landscape, 1860–1861.
L MSS MAR

72. MATTINGLY, Peter Frederick (1914–1993)

Mattingly worked on the staff of the Museum as a mosquito taxonomist from 1946 until he retired in 1979. He published 180 papers, mostly on mosquitoes and their taxonomy.

Biographical Reference: personal information.

Collection

(1) One folder of letters: correspondence concerning the proposal to publish the papers of Charles Darwin held at Down House, 14 letters, 1972–1973. Presented by Dr Mattingly.
L MSS MATT

73. MELVILLE, Richard Valentine (1914–1993)

Born at Tidworth and educated at Wellington College, Melville read geology at Reading University. He was on the staff of the Geological Survey from 1938 until his retirement in 1974, serving as Assistant Director (Specialist Services) for his last two years. Melville was closely associated with the work of the International Commission on Zoological Nomenclature, and was Secretary and Editor of the Palaeontographical Society from 1948 to 1961.

Biographical Reference: 1993. Obituary. *Bulletin of Zoological Nomenclature* 50:101–102.

Collection

(1) One Ms in a folder: *Natural History Books in the Méjanes Library, Aix-en-Provence*, [1968]. Presented by R. V. Melville, 1985.
L MSS MEL

74. MENZIES, Archibald (1754–1842)

Born in Perthshire, Menzies worked as a gardener in the Edinburgh Botanic Gardens, before training as a surgeon in Edinburgh and joining the Royal Navy. He served on the *Nonsuch*, the *Prince of Wales*, and then under Captain George Vancouver on the *Discovery*. Menzies made important botanical collections in all the countries visited by the *Discovery*, particularly in Chile and Canada. His final service was on the *Sanspareil* in the West Indies under Lord Hugh Seymour.

Biographical Reference: *DNB*.

Collection

(1) One folder of Mss: zoological notes made during Vancouver's voyage to the Pacific Ocean, 1790–1795. The notes are on mammals and birds, and include descriptions of the sea otter and American mink. One leaf is headed 'Trinidade, May 3d 93'.
L MSS MEN

75. MONTAGU, George (1755–1815)

Montagu was born in Wiltshire, entered the army at an early age, and served as a Captain in the 15th Regiment of Foot during the American War of Independence. He settled in Tedbury after leaving the army and devoted himself to the study of natural history, and especially birds. He published *Ornithological Dictionary, or Alphabetical Synopsis of British Birds* in 1802.

Biographical Reference: *DNB*.

Collections

(1) One volume in a box: notebook containing natural history observations, mostly written as diary entries, 1796–1798. Presented by Dr H. E. Quick, 1961; previously owned by the Hellins family.
L MSS MON A

(2) One box of Mss: correspondence and papers relating to the Montagu,

Dorville and Wolff families, including wills, death and burial certificates, 'instructions for my executor', papers on nature printing and portraits, 1814–1909; 1965. A finding aid is available. Presented by Miss H. H. Hellins, 1985.
L MSS MON B

(3) One Ts in a folder: *List of the Manuscripts Relating to George Montagu in the Library of the Linnean Society*, compiled by R. J. Cleevely, [c.1975].
L MSS MON C

76. **MUMFORD, Alfred Alexander (1862–1943)**

Mumford worked as the Medical Officer at Manchester Grammar School from 1909 to 1928, and was the author of *Manchester Grammar School, 1515–1915* (1919).

Biographical Reference: personal information.

Collection

(1) One folder: two letters to L. J. Fletcher about the John Abbott drawings, 1917.
L MSS MUM

77. **MUNFORD, Revd George (c.1794–1871)**

Born in Great Yarmouth and educated at Oxford University, where he matriculated in 1818, Munford served as Vicar of East Winch, Norfolk, from 1849 until his death. Munford, who changed his name to Montford late in life, was an amateur botanist and naturalist, and contributed papers to local journals and magazines.

Biographical Reference: 1872. Obituary. *Transactions of the Norfolk and Norwich Naturalists' Society, 1871–1872*, 12.

Collection

(1) Two Ms volumes: *Miscellaneous Notes on Pliny's Natural History, Prepared for the Wernerian Club Edition of the Translation into English by Philemon Holland*, 1847–1851; notebook on natural history with extracts of books read by Munford, 1854–1861. Presented by I. S. Montford, 1955.
L MSS MUN

78. MUSEUM DARWINIANUM, Moscow

The Museum was founded in 1905 as the private collection of Alexander Eric Kohts, and was presented to the Moscow Higher Course for Women ('the women's university') in 1913. Kohts intended the Museum to present 'a visual illustration of Darwin's theory of the evolution of the living world'. The exhibits consisted of stuffed animals arranged to demonstrate variation or aspects of inheritance, paintings, busts, statues and models.

Historical Reference: Kohts, A. 1955. The Museum Darwinianum. *Voks Bulletin* 6(95):26–32.

Collection

(1) Eight photograph albums and folders: photographs of paintings, sculpture and natural history specimens, educational activities, portraits of staff and pictures of the building, [c.1930–1963]. One folder is entitled *Variation in the Plumage of some Tetraoninae, a Contribution to Darwinism, Photographs from Specimens of the Museum*, 1956. Presented by Professor A. E. Kohts, 1959–1963.
L MSS MUS

79. NORMAN, Revd Alfred Merle (1831–1918)

Born in Exeter and educated at Ilminster, Winchester College and Oxford University, Norman studied theology at Wells and was ordained as a priest in 1857. After a number of curacies, Norman was Rector of Burnmoor from 1866 to 1895. He was interested in botany and geology as a youth, but worked on aspects of marine biology in later life. He built up a fine private collection, much of which was acquired by the Museum in 1898. He was the author, with G. S. Brady (1832–1921), of *A Monograph of the British Entomostraca* (1867).

Biographical Reference: 1919. Obituary. *Proceedings of the Royal Society of London*, series B, 90:xlvi–xlix. Portrait.

Collection

(1) One folder of Mss: *Within the Veil*, notes for a sermon on a text from the Book of Leviticus, 1886.
L MSS NOR

80. **OATES, Eugene William (1845–1911)**

Born in Sicily and educated in Bath and by private tutors, Oates joined the Public Works Department of the Indian civil service in 1867. He served in Burma and elsewhere until his retirement in 1899. Oates was an accomplished ornithologist, and published *Game Birds of India* in 1898–1899. He worked on a catalogue of the eggs in The Natural History Museum from 1898 to 1902, and his collection of Burmese birds and their eggs was acquired by the Museum between 1882 and his death.

Biographical Reference: 1912. Obituary. *Ibis*, 9th series, 6:341–342.

Collection

(1) One folder: four letters to R. G. W. Ramsay, 1882.
L MSS OAT

81. **ORDE-POWLETT, William Thomas, 4th Baron Bolton (1845–1922)**

Orde-Powlett was educated at Eton and Trinity College, Cambridge. He lived at Bolton Hall, Leyburn, Yorkshire, and was a Justice of the Peace, Alderman for North Yorkshire and the Honorary Colonel of the Yorkshire Hussars Imperial Yeomanry.

Biographical Reference: 1929. *Who Was Who, 1916–1928*. 1159pp. London.

Collection

(1) One letterbook: 62 letters to R. Bowdler Sharpe, chiefly about his edition of Gilbert White's *Natural History of Selborne* (1901), 1904–1909. Presented 1936.
L MSS ORD

82. THE OWEN COLLECTION

The Owen Collection comprises the scientific correspondence, manuscripts, drawings and annotated books of Sir Richard Owen (1804–1892) together with some correspondence and papers of William Clift (q.v.), Owen's predecessor at the Royal College of Surgeons, and of Sir Everard Home (1756–1832), John Hunter's brother-in-law and executor. The bulk of the manuscripts were presented by C. Davies Sherborn (q.v.) in 1908, and the annotated books by Miss Emily Owen in 1915. Various items by and about Owen have been added to the collection in recent years.

OWEN, Sir Richard (1804–1892)

Born and educated in Lancaster, Owen trained as a surgeon, and became Assistant Conservator at the Royal College of Surgeons in 1827. He was elected Hunterian Professor of Comparative Anatomy at the college in 1836, and became Joint Conservator with William Clift in 1842. In 1856 Owen was appointed Superintendent of the natural history departments of the British Museum. He masterminded the transfer of the departments to South Kensington, finally retiring in 1883. Owen was the most famous comparative anatomist and vertebrate palaeontologist of his generation.

Biographical Reference: *DSB*.

The Owen Collection is catalogued in detail by J. C. Thackray in 'A catalogue of the correspondence, manuscripts and drawings of Richard Owen, William Clift and Sir Everard Home contained within the Owen Collection at The Natural History Museum', *Richard Owen Commemoration*, Natural History Museum Publications, 1992. The Owen drawings are catalogued by Jean M.

Ingles and Frederick C. Sawyer in 'A catalogue of the Richard Owen collection of palaeontological and zoological drawings in the British Museum (Natural History)', *Bulletin of the British Museum (Natural History),* Historical Series 6(5):109–197.

Items in the summary list which follows are arranged in order of their General Library press mark. Unannotated printed items are omitted from the list.

Collections

(1) Monotremata and Marsupialia, from Todd's *Cyclopaedia,* 1841. Interleaved and annotated.
L OC1

(2) Aves, from Todd's *Cyclopaedia,* 1835. Interleaved and annotated.
L OC3

(3) *Odontography*. Two volumes. London. 1840–1845. Annotated.
L OC6

(4) *Lectures on the Anatomy of the Invertebrates*. Second edition, 1855. Annotated.
L OC8

(5) *British Fossil Mammals and Birds*. London, 1846. Two copies, annotated.
L OC9.1–2

(6) *Anatomy of Vertebrates*. Three volume. London, 1866–1868. Two copies, interleaved and annotated.
L OC10

(7) *Principes d'Ostéologie Comparée*. Paris, 1865. Interleaved and annotated.
L OC11

(8) *Palaeontology*. Second edition. Edinburgh, 1861. Interleaved and annotated.
L OC12

(9) *Lectures on the Anatomy of the Invertebrates*. London, 1843. Annotated.
L OC13

(10) Reports on British fossil reptiles, parts 1 and 2. *Reports of British Association for the Advancement of Science, 1839 [and 1840]*. Interleaved and annotated.
L OC14

(11) *On the Extent and Aims of a National Museum of Natural History*. London, 1862. Annotated.
L OC15.1

(12) *On Parthenogenesis*. London, 1849. Two copies, interleaved and annotated.
L OC17.1

(13) *On the Nature of Limbs*. London, 1849. Two copies, interleaved and annotated.
L OC18

(14) *Experimental Physiology, its Benefits to Mankind*. London, 1882. Sparingly annotated.
L OC19

(15) Diary, 1869 and 1872.
L OC20

(16) Notebook, 1833.
L OC21

(17) Memoir of Professor Owen, from *Men of Eminence*, 1868. Interleaved and annotated.
L OC24

(18) Fifteen pocket notebooks, 1830–1839.
L OC25

(19) *Description of Skeleton of Extinct Giant Sloth*. London, 1842. Interleaved and annotated.
L OC26

(20) *Zoology of the Voyage of the Beagle, Part 1*. London, 1840. Annotated.
L OC27

(21) *Odontography*. Five volumes. London, 1840–1845. Annotated.
L OC28

(22) *Monograph on the Aye-Aye*. London, 1863. Interleaved and annotated.
L OC31

(23) *Fossil Reptilia of South Africa*. Two volumes. London, 1876. Sparingly annotated.
L OC32

(24) *Memoir on the Pearly Nautilus*. Author's Ms drafts.
L OC33.1

(25) *Memoir on the Pearly Nautilus*. London, 1832. Interleaved and annotated.
L OC33.2

(26) Sea serpent scrapbook, 1836–1912.
L OC36

(27) Ms notes and synopses of Hunterian and other lectures, 1828–1864, three volumes.
L OC38

(28) Corrected proofs of reviews by Owen published in *Quarterly Review* and elsewhere, 1847–1885.
L OC39

(29) *Monograph on the Reptilia from the Kimmeridge Clay and Portland Stone*. London, 1869. Annotated.
L OC49

(30) *Observations on Geology* by J. Hunter, edited by Owen, 1859. Annotated.
L OC50

(31) *Memoir on the Megatherium*. London, 1860. Annotated.
L OC54

(32) *Memoir on the Dodo*. London, 1866. Interleaved and annotated.
L OC55

(33) *Anatomy of the King Crab*. London, 1873. Sparingly annotated.
L OC56

(34) Über die Bedeutung der Schädelknochen, by Oken, 1807. Sparingly annotated.
L OC57

(35) Twenty-eight Mss on natural history by Richard Owen and others. Two volumes.
L OC59

(36) *Memoir on the Extinct Wingless Birds of New Zealand*. Two volumes. London, 1878. Annotated.
L OC61

(37) Correspondence of Richard Owen and William Clift, 27 volumes and three supplementary volumes, c.1725–1892, including T. Allis (11, 1837–1872), D.T. Ansted (12, 1841–1846), G.D. Campbell, 8th Duke of Argyll (32, 1850–1887), R. Ball (17, 1840–1855), M. Barry (26, 1840–1855), S.H. Beckles (12, 1854–1885), T. Bell (47, 1849–1873), G. Bennett (68, 1833–1888), G.F. Bennett (13, 1877–1882), W.H. Bensted (23, 1840–1869), H.M.D. de Blainville (11, 1833–1850), C.C. Blake (13, 1855–1882), E.A. Bond (20, 1879–1885), J.S. Bowerbank (33, 1841–1873), W.J. Broderip (165, 1830–1858), J. Brown (31, 1842–1859), F.T. Buckland (22, 1847–1879), W.B. Carpenter (17, 1842–1854), Sir J. Clark (19, 1837–1886), W. Clift (113, mostly to his wife and sisters, with many drafts of outgoing letters, 1812–1848), H. Coote (11, 1847–1864), T. Craggs (11, 1865–1868), T. Davidson (11, 1852–1878), F. Dixon (28, 1841–1849), Sir P. de M.G. Egerton (58, 1839–1873), W.W. Cole, 3rd Earl of Enniskillen (95, 1835–1885), A.P. Falconer (25, 1846–1873), E. Forbes (12, 1848–1853), W. Fox (15, 1863–1870), J.E. Gray (45, 1842–1873), A.C.L.G. Gunther (27, 1862–1880), Sir J.F.J. von Haast (23, 1859–1886), Barbara, Marchioness of Hastings (64, 1846–1849), T. Hawkins (12, 1838–1851), J. Hector (16, 1859–1880), J.S. Henslow (21, 1843–1859), J. van der Hoeven (23, 1840–1865), G.B. Holmes (23, 1841–1873), T.H.C. Hood (14, 1861–1887), J.W. Jones (15, 1850–1878), T.R. Jones (27, 1851–1883), J.J. Kaup (20, 1836–1873), J.L.G. Krefft (18, 1862–1874), W.G. Lettsom (13, 1864–1883), W.A. Lloyd (14, 1853–1879), Sir C. Lyell (48, 1837–1858), G.A.

Mantell (18, 1840–1848), D.W. Mitchell (44, 1848–1859), C. Moore (11, 1848–1878), Sir R.I. Murchison (60, 1840–1869), A. Newton (23, 1856–1871), Sir A. Panizzi (28, 1856–1866), Sir W. Parish (11, 1836–1856), J.B. Pentland (38, 1834–1855), J. Phillips (16, 1841–1881), E.P. Ramsay (24, 1877–1889), G. Ransome (20, 1847–1856), L. Reeve (16, 1842–1865), Sir J. Richardson (11, 1841–1858), G.D. Rowley (15, 1866–1878), Sir E. Sabine (14, 1851–1870), P.L. Sclater (16, 1859–1881), A. Sedgwick (40, 1842–1869), J.D. Dunbar, 6th Earl of Selkirk (17, 1853–1884), W. Sharpey (11, 1842–1864), C.H. Smith (12, 1840–1855), J.T. Smith (31, 1847–1866), Sir G.G. Stokes (18, 1865–1889), S. Stutchbury (15, 1836–1858), A. Valenciennes (13, 1835–1863), J. Van Voorst (11, 1843–1861), G.R. Waterhouse (12, 1840–1874), H. Woodward (24, 1879–1888), T. Wright (11, 1854–1882).
L OC62

(38) Mss, lecture notes and correspondence of Sir Everard Home, two volumes, 1782–1832.
L OC63

(39) List of orders and medals of R. Owen (Ts, 1972).
L OC66

(40) Survey of life and work of Owen by D.L. Ross (unpublished Ts, 1972).
L OC67

(41) Armorial bearings of Richard Owen, prepared by The College of Arms, 1973.
L OC69

(42) Drawings and paintings by William Clift and others to illustrate the papers of Sir Everard Home, 1768–1825.
L OC71

(43) *State of the Rooms of the Department of Natural History* [of the British Museum]. Ms, 1824.
L OC73

(44) Diplomas and certificates of Sir Richard Owen.
L OC74

(45) Ms document appointing Daniel Solander to the British Museum, 1765.
L OC76

(46) Ms notes on *Placuna placenta* and *Glyptodon*, 1838 and undated. Purchased 1970.
L OC78

(47) Letter of J. Phillips, 1857, offering Owen presidency of the British Association for 1858.
L OC83

(48) Notes by Owen on his family, undated.
L OC84.1

(49) Music written out by Caroline Amelia Clift.
L OC84.2

(50) An account of the life and achievements of R. Owen by J. Dobson (Ts, 1981).
L OC86

(51) The Richard Owen correspondence – a calendar and an introductory essay by J. Gruber (Ts).
L OC87

(52) Portraits of Richard Owen, mounted in an album.
L OC88

(53) Drawings of the nervous system of bivalves, ?1834.
L OC89

(54) Scientific Mss of Richard Owen, newscuttings and other items, 1830–1878, four volumes.
L OC90

(55) The Owen collection of palaeontological and zoological drawings, mounted on 523 sheets, including W. Clift (35), J. Dinkel (42), J. Erxleben (28), J. J. Kaup (13), R. Owen (83), G. Scharf sen. and jun. (35) and H. Scharf (23).
L OC91

(56) Photocopies of Owen letters in the American Philosophical Society.
L OC96

(57) Eighty-two woodblocks intended to illustrate 'Fossil Osteology'. Presented by J. Murray, 1921.
L OC97

(58) Reply to criticisms by a Royal Society Referee, 1870.
L OC98

(59) On the anatomy of the American kingcrab, 1871.
L OC99

(60) Fossil Crocodilia of the Wealden and Purbeck formations, c.1879. Presented by Professor B. G. Gardner, 1993.
L OC103

(61) Signatures of those attending the Red Lions Dinner, Aberdeen, 19 September 1859. Presented by Professor B. G. Gardner, 1993.
L OC104

83. PERCIVAL, Arthur Blayney (1875–1941)

Percival was born in Newcastle upon Tyne, and educated there and at Durham College of Science. He visited Transvaal and Nyasaland as a young man, before leading an expedition to southern Arabia for Lord Rothschild and The Natural History Museum in 1899. He worked as a Ranger in the Game Department of the Kenyan government, from 1902 until he retired in 1928. He presented a collection of birds from Kenya to the Museum in 1909.

Biographical Reference: 1941. Obituary. *Ibis,* 14th series, 5:468–470.

Collection

(1) One Ms volume: *Notes on an Expedition into South Western Arabia,* with a sketch map, 1899. Presented by the author, 1899.
L MSS PER

84. PERCY SLADEN Expedition to Palestine, 1935

The Percy Sladen Fund financed a two-man expedition, comprising Mr Roger Washbourn and Mr R. F. Jones, both of the University of Birmingham, to carry out a biological survey of Lake Huleh. The lake, which lay at the northern end of the Jordan Valley, was shortly to be drained by the Palestine Land Development Co.

Historical Reference: Anon. 1935. The Percy Sladen Expedition to Lake Huleh. *Nature, London* 136:538.

Collection

(1) One folder of Mss: notebook of R. Washbourn, being the expedition diary for August to December 1935; two chemical analyses of lake water, 1936; a brief typed report on the results of the expedition, [1936].
L MSS PERCY

85. PITT, Frances (1888–1964)

Frances Pitt was a student of animal psychology and behaviour, and a prolific writer on British wildlife and garden natural history. A popular lecturer, she illustrated her talks with her own films.

Biographical Reference: 1972. *Who Was Who, 1961–1970*. 1243pp. London.

Collection

(1) Four boxes of films: 67 reels of 16mm film of natural history subjects in England, Wales and Scotland, Holland, Iceland, Ireland, Lapland, Norway; the United States and Canada; Egypt, Kenya and Uganda, 1939–1964. Presented by the National Trust, 1988.
L MSS PIT

86. PLANT, John (1819–1894)

Born and educated in Leicester, Plant worked in his father's business. He became Honorary Secretary of the Leicester Naturalists' Club, and Curator of the Museum of the Literary and Philosophical Society. In 1849 he moved to Salford to become Curator and Chief Librarian at the Peel Park Museum. Plant was particularly interested in geology and palaeontology, and was a Fellow of the Geological Society of London.

Biographical Reference: General Library Handwriting Collection, The Natural History Museum.

Collection

(1) One folder: seven letters from naturalists, including James Harley (3), 1847.
L MSS PLAN

87. PORT JACKSON PAINTER

The 'Port Jackson Painter' is a generic term applied to one or more unknown watercolour artists working in Sydney from 1788 through to the 1790s, and who adopted a particular style of painting in which certain features, such as rock markings, were represented in a stylized convention.

Biographical Reference: 1992. *Dictionary of Australian Artists*.

Collection

(1) Seven boxes: 70 watercolour drawings of aborigines, animals and plants made near Port Jackson, [1788–1800]. Formerly in the collection of Sir Joseph Banks (Ms 34).
L MSS POR

88. PULTENEY Richard (1730–1801)

Born and educated in Loughborough, Pulteney practised as a surgeon and apothecary in Leicester. He graduated MD in Edinburgh in 1764 and worked as a physician in Blandford, Dorset, for the rest of his life. He amassed a fine library and a museum of shells, minerals and dried plants. Pulteney was author of *Historical and Biographical Sketch of the Progress of Botany in England* (1790).

Biographical Reference: *DNB*.

Collections

(1) One letterbook: correspondence, including E. M. Da Costa (12, 1776–1778, with copies of Pulteney's replies, lists etc.), G. Montagu (4, 1800, with copies of replies, and a catalogue) and H. Seymer (13, 1781–1783, with copies of replies). Transferred from the Department of Zoology, 1939.
L MSS PUL A

(2) One letter in a folder: one letter from William Hanbury, 1788. Purchased 1993.
L MSS PUL B

89. RAY SOCIETY

Founded in 1844 with the object of promoting natural history by printing original works in zoology and botany, new editions of works of established merit, rare tracts and manuscripts, and translations and reprints of foreign works which are generally inaccessible from the language in which they are written, or from the manner in which they have been published.

Historical Reference: Curle, R. 1954. *The Ray Society, a Bibliographical History*. 101pp. London.

Collections

The Ray Society Archive was presented to the Museum in 1985.

(1) Five Ms volumes: *Minutes of Meetings*, 1844–1847, 1859–1948. The Ms minutes of Council meetings, signed by the president, are bound with printed notices of meetings, annual reports, rules, memoranda and letters.
L MSS RAY A

(2) Two Ms volumes: *Minutes of Meetings*, 1855–1898. Rough minutes of Council meetings, from which the formal record was prepared.
L MSS RAY B

(3) Two Ms volumes: *Meetings Signature Book*, 1876–1973. Contains signatures of Council members attending their meetings.
L MSS RAY C

(4) One Ms volume: *Reports of Council*, 1874–1966. Printed reports submitted to members at the Annual General Meeting. Other copies are bound into the Council minute books.
L MSS RAY D

(5) Eight Ms volumes: *Cash Books*, 1854–1864, 1905–1961. A record of payments made to and by the Society.
L MSS RAY E

(6) Five Ms volumes: *Subscription Ledger*, 1856–1889, 1904–1966. Contains a record of payments made by individual and institutional subscribers, with letters, notes, newscuttings and other documents.
L MSS RAY F

(7) One printed/Ms volume: *Ray Society*, 1848–1861. Contains printed lists of members, lists of publications, reports, rules and notices of meetings, Ms minutes for two Council meetings in 1856, and duplicated minutes for 1848–1856.
L MSS RAY G

(8) One letterbook: *Jardine Hall Mss, Vol. XVII, Ray Society Letters*, 1843–1844. Letters concerning the foundation of the Society addressed to G.

Johnston and Sir W. Jardine, together with a Ms entitled *Reality, a Tale of Physical Science Founded on Facts* by [Sir W. Jardine]. Presented to the Ray Society by N. B. Kinnear, 1920.
L MSS RAY H

90. **READE, Hester Carter** (1785/6–1874)

Hester Reade was an accomplished amateur naturalist who lived in Camberwell and later in Hastings. Probably a schoolteacher, she was the sister-in-law of the Revd G. E. Smith, and several of her drawings appear in Smith's scrapbook (q.v.).

Biographical Reference: internal evidence.

Collection

(1) One small Ms volume: *Book of Reference to The Medley, Game of Choice. For the Instruction and Amusement of Younger Pupils,* Camberwell, 1822. There is a separate card giving instructions on playing the game. Presented by Miss A. Clark, 1933.
L MSS REA

91. **RICHARDSON, Richard** (1663–1741)

Born at North Bierley, Yorkshire, Richardson was educated at Bradford and Oxford University, and obtained an MD from the University of Leyden. He made numerous botanical expeditions through the British Isles, and had a fine garden at his home in North Bierley. Richardson was a Fellow of the Royal Society, and a respected naturalist and antiquary.

Biographical Reference: *DNB*.

Collection

(1) One letter in a folder: one letter to Philip Miller, Chelsea, 1727. Purchased 1979.
L MSS RIC

92. ROSEVEAR, Donovan Reginald (1900–1986)

Born in Chelsea, and educated at Bedford and Cambridge University, Rosevear joined the Colonial Forest Service in 1924. He worked in Nigeria, becoming Inspector General of Forests in 1951, and retired in 1954. Rosevear worked as an Honorary Associate at The Natural History Museum, preparing a series of monographs on the mammals of West Africa, which were published between 1965 and 1974.

Biographical Reference: 1988. Obituary. *The Nigerian Field* 50:51–52.

Collection

(1) Four volumes and two boxes of Mss: an autobiography with notes and reminiscences, natural history Mss and Tss, photographs of homes in Nigeria, correspondence and diaries, and reviews of his published works, 1922–1986. A finding aid is available. Presented by Mrs G-M. Rosevear, 1990.
L MSS ROS

93. ROYAL SOCIETY OF LONDON

Founded in 1660 as The Society for the Promoting of Physico-Mathematicall Philosophy, it is Britain's premier scientific society.

Historical Reference: Hall, M. B. 1984. *All Scientists Now, the Royal Society in the Nineteenth Century*. 262pp. Cambridge: Cambridge University Press.

Collection

(1) Two volumes and one box of Mss: letters addressed to Dr A. Gunther concerning publication of 'An account of the petrological, botanical and zoological collections made in Kerguelen's Land and Rodriguez during the Transit of Venus expeditions carried out by order of Her Majesty's Government in the years 1874–75', *Philosophical Transactions of the Royal Society* 168, 1879, including I. B. Balfour (16, 1877–1878), J. W. Clark (17, 1876–1879), A. E.

Eaton (18, 1876–1878), and W. White (14, 1876–1879), with authors' Mss of many of the published papers. A finding aid is available.
L MSS ROY

94. SALE NATURAL HISTORY SOCIETY

Founded in 1883 with Francis Amos as President and R. J. W. C. McBeath as the Secretary. Meetings were held in the Congregational Sunday School in Montague Road, Sale. The Society survived until 1886.

No published references.

Collection

(1) One Ms volume: *Papers Read Before the Sale Natural History Society, Volume 1,* 1884–1885. Contains an introduction by the President, Francis Amos, and 60 papers by 15 authors read between February 1884 and February 1885. Purchased 1949.
L MSS SALE

95. SAUNDERS, William Wilson (1809–1879)

Born in Wendover and educated privately, Saunders worked for Lloyds of London. He lived in Wandsworth and, from 1857, in Reigate. He was an enthusiastic entomologist and botanist, and founded the Holmesdale Natural History Club in 1857. He donated and sold collections of plants and insects to the Museum between 1865 and his death.

Biographical reference: *DNB*.

Collection

(1) Three volumes of Mss: notes on the natural history of Wandsworth, Reigate and elsewhere, with pencil and watercolour sketches, 1848–1869. The volumes bear the signature of George Sharp Saunders (1842–1910). Acquired in 1928.
L MSS SAU

96. SCIENCE GOSSIP (1865–1902)

A popular magazine published by Robert Hardwicke, and later Chatto & Windus. The series consisted of volumes 1–29, 1865–1893, and New Series volumes 1–8, 1894–1902.

Collection

(1) Two volumes of drawings: small pen-and-ink and pencil drawings, and photographs used a sources for the illustrations published in *Science Gossip*, together with printed proofs, c.1865–1900.
L MSS SCI

97. SCOTT, Hugh (1885–1960)

Scott studied natural science at Trinity College, Cambridge. In 1908 he joined the Second Percy Sladen Trust Expedition, studying the fauna of the Seychelles, and was the general editor of the insect sections of the Seychelles reports. On his return to England in 1909 he became the Curator of Entomology at Cambridge University Museum, a post he held until 1928. In 1930 he became Assistant Keeper in the Department of Entomology at The Natural History Museum, and remained in post until his retirement in 1948. He published many scientific and geographical papers, including *In the High Yemen* (1942).

Biographical Reference: 1960. Obituary. *Entomologist's Monthly Magazine* 96:105.

Collections

(1) Twelve Ms volumes and one box: diaries and journals, covering Scott's home life in and around Oxford, and his travels in Scandinavia, Italy, Spain, Iraq, the Seychelles, West Indies, Algeria, South Africa and elsewhere, 1899–1937. The journals have inserted sketches, newscuttings, pressed flowers, photographs and printed ephemera.
L MSS SCO A

(2) One folder of Mss: two journals relating to holidays in the Canary Islands, 1955–1959; notes relating to visits to Abyssinia in 1926, 1948 and 1952.
L MSS SCO B

(3) One folder: letters from Hugh Scott to members of his family, written from Abyssinia, 1926–1927.
L MSS SCO C

(4) Seven photograph albums with one folder of Mss: mounted and captioned photographs and lists, recording expeditions to Aden, Yemen and the West Aden Protectorate, 1937–1938; the Gughe Highlands of southern Ethiopia, 1948–1949; and northern Ethiopia, 1952–1953.
L MSS SCO D

(5) One Ms volume: *Log of the Sixth Anglo-Catholic Pilgrimage to Palestine, Constantinople, Athens etc.*, by the Revd Prebendary Ernest Scott (1892–1949), the only brother of Hugh Scott, 1930. The volume has inserted maps, photographs and newscuttings. Presented by Dr Hugh Scott, 1957.
L MSS SCO E

98. **SHARPE, Richard Bowdler (1847–1909)**

Born in London and educated in Peterborough and Loughborough, Sharpe worked as a clerk, bookseller's assistant and a librarian from 1863 to 1872. He had studied birds from an early age, and came to the notice of J. E. Gray (q.v.), Keeper of Zoology at the British Museum, who obtained a post for him at the Museum. Sharpe took charge of the bird collection at the Museum, and devoted the rest of his life to building up and cataloguing the collection. He was author of *A Hand-list of the Genera and Species of Birds* (1899–1909).

Biographical Reference: *DNB*.

Collection

(1) One folder: five letters to A. Hay, 9th Marquis of Tweeddale, about birds, 1874–1878.
L MSS SHA

99. **SHAW, George (1751–1813)**

Shaw was born in Brierton, Buckinghamshire, and educated at home and at Oxford. He was ordained a deacon in 1774, but his enthusiasm for natural history took him away from the ministry. He studied medicine in Edinburgh and entered into practice in London in 1787. Shaw became an Assistant Keeper in the Natural History Department of the British Museum, and was Keeper from 1807 until his death. He was a voluminous writer on natural history, and his publications include *The Naturalist's Miscellany* (24 volumes, 1789–1813).

Biographical Reference: DNB.

Collection

(1) One volume of drawings: 185 pencil, pen-and-ink and watercolour sketches of animals, plants and mineral specimens, many of them published in *General Zoology* (1800–1826) and other works by Shaw, with autograph notes. Presented in June 1896.
L MSS SHAW

100. **SHERBORN, Charles Davies (1861–1942)**

Sherborn was an enthusiastic geologist as a young man, carrying out research on microfossils and the geology of south-east England. He became convinced of the importance of bibliographical and historical research as the foundation of taxonomy, and spent over forty years compiling *Index Animalium* (1902–1933), which comprised a complete list of the generic and specific names published between 1758 and 1850, with a reference to the date and place of publication of each name. The General Library of The Natural History Museum was his base, although he was never a salaried member of staff.

Biographical Reference: Norman, J. R. 1944. *Squire. Memories of Charles Davies Sherborn*. 202pp. London.

Collections

(1) One folder of Mss: nine letters to Sherborn concerning his *Index Animalium*, 1889–1892; *Details of Scheme of 'Index Generum et Specierum Animalium'*, drawn up by Sherborn for the Trustees of The Natural History Museum, 1891; *Index Animalium. State of Manuscript, October 1924*, with notes updating the statement to June 1927. Presented by C. D. Sherborn.
L MSS SHE A

(2) One volume of Mss: *'Index Animalium' Ms notes*, 1920–1928. This collection comprises notes and extracts from rare publications sent to Sherborn by librarians and naturalists in France, Germany, the United States and elsewhere. Presented by C. D. Sherborn.
L MSS SHE A

(3) One volume of Mss: *Ledger*, 1890–c.1940. The volume contains notes on the progress of the *Index Animalium*, slips for books not seen, lists of books purchased by Sherborn, lists of desiderata for the Museum libraries, and lists of photostats in the Museum libraries. Presented by C. D. Sherborn.
L MSS SHE A

(4) One printed volume, annotated: *Where is the – Collection?*, 1940. The volume contains typed and printed insertions as well as annotations by Sherborn and others. Presented by C. D. Sherborn.
L MSS SHE A

(5) One folder of Mss and printed ephemera: *Broadside, Cartoons etc. E Coll Sherborn*, 1772–1915. 73 items mounted on 56 sheets, including advertisements for animal shows, the printed ephemera of learned societies, poems, admittance tickets for menageries, cartoons, caricatures and trade cards. Many items are from the Owen – Clift collection. Presented by C. D. Sherborn, 1931.
L MSS SHE A

(6) One Ms volume: *Copy of a Paper on Living in the Tropics by the Late Charles Waterton Esq.*, transcribed by Norman Moore, 1867, from a Ms dated 1838. Presented by C. D. Sherborn, 1933.
L MSS SHE A

A Noble Lioness
And her Three fine Young Cubs,
ARE NOW TO BE SEEN AT THE
ROYAL MENAGERIE, EXETER 'CHANGE.
Indisputably the Grandest Assemblage of Living Curiosities in the known World.

In the same Apartment—Admittance One Shilling,
TWO of the most extraordinary FULL-GROWN
MALE LIONS,
Ever seen in this Kingdom, having fine large flowing Manes; and a famous
LIONESS.
Of the same Class, which instantly impresses every Beholder with that magnificent Idea so generally conceived of the King of the Brute Creation. The Form of these Lions are strikingly bold and majestic, having large shaggy Manes, which they can erect at Pleasure, and their tremendous Roaring, like Thunder, exhibits a Picture of Grandeur, which no Words can describe.
NOBLE
Bengal Royal Tiger.
The Ferocious PANTHER, from Buenos Ayres.
ELEGANT LEOPARDS,
FROM ALGIERS AND CEYLON,
FOUR LAUGHING HYÆNAS,
From the Cape of Good Hope.
A striped Abyssinian Hyæna,
Two large Oriental Porcupines,
From CEYLON, covered with an immense Number of hard spiny Quills, many of which are 20 Inches long.
THREE FIERY LYNXES,
The most Ferocious and Watchful of all Animals.
A GREAT WHITE POLAR,
Generally exhibited under the Title of the SEA LION, the most tremendous Animal of
THE FROZEN OCEAN,
With an extensive and pleasing Variety of the Simai Tribe, and other curious and interesting Animals.

Second Apartment, Admittance One Shilling,
IS THE LARGEST ANIMAL EVER SEEN IN EUROPE, A MALE
ELEPHANT
Ten Feet high, and upwards of Four Tons weight,
By far exceeding in Height and Bulk any other Animal ever seen in England, with large Ivory Tusks, one standing out on each Side of his Trunk; his Sagacity is beyond Conception, he performs many very curious Tricks, and his Consumption in Hay Corn, Straw, Carrots, Water, &c. exceeds Seven Hundred Pounds daily.

100(5) Sherborn, Charles Davies (1861–1942)
A broadsheet advertising the animals on view at the Royal Menagerie, Exeter 'Change, in the Strand, London. The menagerie remained at this address from its foundation in 1794 until 1829, when it moved to Charing Cross.

(7) One box of slips: index slips for a catalogue of the type and figured specimens described in *Mineral Conchology* by J. and J. de C. Sowerby, undated. Presented by C. D. Sherborn.
L MSS SHE A

(8) Four boxes and one folder of Mss: c.300 autograph letters of naturalists and others, selected by Sherborn from his own correspondence, and those of C. W. Andrews, Mr Banting, T. D. A. Cockerell, T. R. Jones, A. W. Rowe, A. S. Woodward and others, c.1840–1935; eight notebooks, address books, lists of desiderata and books acquired, c.1900–1935; 13 pamphlets and two booklets presented to Sherborn, 1898–1910; printed ephemera, Ms notes and sketches, and a copy of Sherborn's will, 1839–1944; photographs of family and friends, scientific colleagues, his house in Fulham etc., c.1880–1940; dissected 1-inch-to-the-mile Ordnance Survey maps of parts of Great Britain, some with geological annotations and colouring, c.1880–1900; bookplate and view etched by C. W. Sherborn; printed admission tickets for libraries and museums; signatures of Sherborn's friends and colleagues; letter of congratulation from C. Tate Regan on completion of *Index Animalium*, 1890–1933. A finding aid is available. All presented by W. H. T. Tams, 1973–1985.
L MSS SHE B

(9) One Ms volume: *Index to Autograph Letters, or Handwritings, of Persons Interested in Science, Collected by Charles Davies Sherborn, and Preserved . . . in the General Library of the British Museum (Natural History)*, undated. Purchased 1986.
L MSS SHE C

101. **SIBREE, Revd James (1836–1928)**

Born in Hull, Sibree trained as an architect. He was ordained in 1867, and spent the next thirty-eight years as a missionary in Madagascar. Sibree became an expert on the natural history and anthropology of the island, and published a number of papers on the birds and mammals of Madagascar.

Biographical Reference: 1929. Obituary. *Ibis*, 12th series, 5:703–704.

Collection

(1) One volume of Mss and printed cuttings: *Madagascar Mammals and other Natural History Notes*, 1892–1928. Presented by Miss E. Sibree, 1930.
L MSS SIB

102. **SIMPSON, Norman Douglas (1890–1974)**

Douglas Simpson was born in Yorkshire and educated at Clifton College, Bristol, and Trinity College, Cambridge. He was already a keen field botanist by the age of 12 and, after service in the First World War, worked as an economic botanist in Egypt and later Ceylon. Simpson retired at the age of 43, and spent the rest of his life annotating and arranging his herbarium, building up his library, and compiling *A Bibliographical Index of the British Flora* (1960). After his death, Simpson's collections were divided between the Botany School, Cambridge, the Royal Botanic Gardens, Kew, and The Natural History Museum.

Biographical Reference: 1975. Obituary. *Watsonia* 10:403–410.

Collection

(1) Four boxes and one folder of Mss: notes and correspondence relating to *A Bibliographical Index*, arranged topographically, 1928–1958, including J. Gibbons (10, 1948–1958), J. D. Grose (27, 1945–1958, and D. H. Kent (17, 1947–1954); papers and correspondence on the publication of *A Bibliographical Index*, 1945–1962, including D. H. Kent (11, 1957–1961); botanical notebooks, papers and correspondence, 1885–1974, including J. S. L. Gilmour (33, 1942–1948); printed ephemera of the Botanical Society of the British Isles, 1935–1970; five watercolour drawings, undated. Presented by N. D. Simpson's executors, 1974.
L MSS SIM

103. **SINEL, Joseph (1844–1929)**

Born on Jersey, Sinel developed an early interest in chemistry and natural history. He worked in retail trade until 1883, when he resigned to become a taxidermist and dealer in natural history specimens. He studied and wrote on the natural history, geology and archaeology of Jersey and Guernsey, and was instrumental in setting up the Jersey Biological Sation in 1891. Sinel was author of *An Outline of the Natural History of our Shores* (1906).

Biographical Reference: 1931. *Report and Transactions of the Société Guernesiaise (1929)*: 369–371.

Collection

(1) One volume of Mss: *Zoological Preparations for Sale,* being a priced list and details of collections sent out, 1889–1890.
L MSS SIN

104. **SMITH, Edward**

Born in Dunstable, Smith was a bibliographer, antiquarian and writer on historical topics, and author of *The Life of Sir Joseph Banks* (1911).

Biographical Reference: 1929. *Who Was Who, 1916–1928*. 1159pp. London.

Collection

(1) One Ms in a folder: calendar [of some letters of Sir Joseph Banks], undated. Presented by W. R. Dawson.
L MSS SMI

105. **SMITH, Revd Gerard Edwards (1804–1881)**

Born in London and educated at Merchant Taylors School and St John's College, Oxford, Smith obtained his BA in 1829. He took holy orders and served as a vicar in parishes in Sussex, Yorkshire and Derbyshire. He was a keen botanist and published a *Catalogue of Plants Collected in South Kent* (1829).

Biographical Reference: *DNB*.

Collection

(1) One scrapbook: *Missionary Needlework etc.*, a volume of natural history drawings, notes, newscuttings, printed illustrations and photographs, arranged systematically, 1792–1879. Many of the notes and drawings are by H. C. Reade (q.v.). Presented by Miss A. Clark, 1933.
L MSS SMIT

106. THE SOWERBY COLLECTION

The Sowerby Collection consists of the papers and correspondence of three generations of the Sowerby family of natural history illustrators and engravers. A finding aid is available. The bulk of the collection was purchased at auction in 1969, with further purchases from dealers and private individuals between 1970 and 1980. The papers of J. W. Salter were presented by Mrs M. Miller in 1985.
L MSS SOWERBY COLL

SOWERBY, James (1757–1822)

Born in London, James's father died young, and he was apprenticed to a painter in 1771. He moved from portraiture to landscape, and eventually to botanical painting. James Sowerby married Anne de Carle in 1786 and set up house in Lambeth. His first independent work was *Flora Luxurians* (1789–1791), with *A Botanical Drawing Book* (1790) following soon after. He collaborated with James Edward Smith (1759–1828) on the illustrations for eight botanical books between 1789 and 1814, and produced books on minerals and fossils between 1811 and the time of his death.

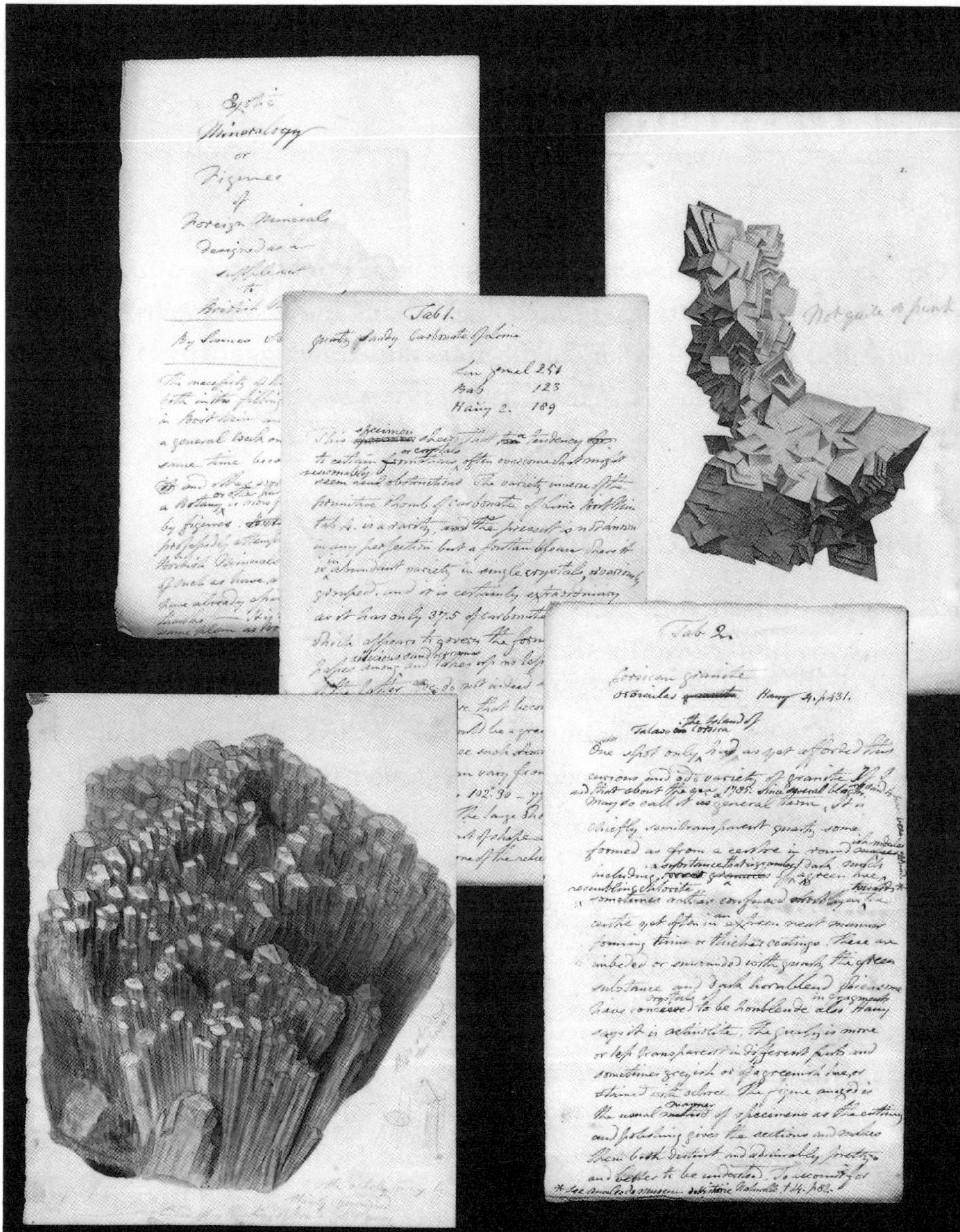

106(2) Sowerby Collection
Draft text, a watercolour drawing and a proof engraving for *Exotic Mineralogy*, a work by James Sowerby (1757–1822), published in twenty-seven parts between 1811 and 1820.

SOWERBY, James de Carle (1787–1871)

Born in London, the eldest son of James (q.v.) and Anne Sowerby, James de Carle became interested in chemistry and mineral analysis as a young man. He drew illustrations for Dawson Turner (q.v.) in 1804, but did not add his name to any engraved plates until after his father's death in 1822. J. de C. Sowerby went into partnership with his younger brother Charles at this date, and moved to Camden Town with his wife and eight children when the partnership was dissolved in 1831. He specialized in zoological and fossil subjects, and undertook large numbers of plates for *Annals and Magazine of Natural History*. Sowerby was Secretary of the Royal Botanic Society from 1838 until his death.

SOWERBY, William (1827–1906)

William, the second son of James de Carle (q.v.) and Anne, followed his father as an artist and engraver, although his output was much smaller. He succeeded his father as Secretary of the Royal Botanic Society, and lived in Regents Park all his life. He married his cousin Charlotte Bryant about 1854 and had four children. He resigned his post as Secretary in 1895 and was succeeded by his son James Bryant Sowerby.

SALTER, John William (1820–1869)

Salter was bound apprentice to James de Carle Sowerby (q.v.) in 1835, and lived with the family in Camden Town and later Regents Park. He worked as artist and engraver, before obtaining a post with the Geological Survey of Great Britain in 1846. He married Sarah (known as Sally) Sowerby, his master's daughter, in the same year.

Biographical Reference: 1974. Papers on the Sowerby family. *Journal of the Society for the Bibliography of Natural History* 6(6): 373–568.

Collections

(1) Twenty-four boxes of letterbooks: letters to members of the Sowerby family arranged alphabetically by correspondent, 1795–1910, including C. Abbot (64, 1795–1812), Lady M. E. Arden (16, 1795–1837), C. C. Babington (10, 1828–1844), W. L. Baker (10, 1798–1816), T. Bell (165, 1825–1834 and undated), M. J. Berkeley (19, 1832–1865), W. Bingley (34, 1799–1814), W. Borrer (128, 1801–1861), J. S. Bowerbank (30, 1834–1871), E. W. Brayley (19, 1818–1840), J. Brodie (25, 1795–1816), R. Brown (16, 1807–1833), S. P. Bryer (11, 1800–1802), W. Buckland (55, 1833–1835), E. Charlesworth (10, 1835–1840), E. D. Clarke (13, 1804–1819), Miss J. C. Codrington (12, 1805–1822), J. Dalton (25, 1802–1815), H. Davies (38, 1794–1818), Revd J. Davies (10, 1796–1815), Lord Dehaldan (13, 1857–1865), W. H. Fitton (74, 1827–1840), T. Frankland (13, 1798–1800), W. Gassiot (10, 1871–1889), C. de Gerville (13, 1814–1826), J. E. Gray (10, 1827–1874), M. L. Green (10, 1885–1886), Revd J. Harriman (46, 1798–1807), J. Harris (10, 1835–1836), A. H. Hassell (10, 1842), Revd J. Hemsted (34, 1795–1812), Revd J. Holme (17, 1800–1812), W. J. Hooker (31, 1804–1865), A. B. Lambert (18, 1795–1825), W. E. Leach (17, 1809–1826), J. Lindley (26, 1818–1840), Longman & Co. (37, 1806–1840), W. Lonsdale (24, 1820–1849), J. C. Loudon (52, 1821–1842), C. Lyell sen. (17, 1806–1812), C. Lyell jun. (10, 1825–1840), Revd T. O. Marsh (13, 1797–1814), J. Mawe (16, 1814–1822), A. McLeay (11, 1804–1819), T. Meade (18, 1803–1829), St G. J. Mivart (10, 1880–1885), G. Montagu (29, 1801–1806), W. J. T. Moreton, Lord Ducie (10, 1837–1844), R. I. Murchison (131, 1827–1861), J. G. Murray (23, 1798–1808), G. H. Noehden (35, 1798–1821), T. Purton (34, 1812–1830), B. Quaritch & Co. (16, 1869–1878), P. Rashleigh (22, 1803–1808), Revd R. Relhan (14, 1793–1806), E. Robson (16, 1795–1802), M. Royer (10, 1820–1821), R. Salisbury (11, 1790–1800), J. Salt (11, 1797–1808), S. W. Silver (101, 1861–1878), J. Sims (25, 1800–1818), Revd G. E. Smith (14, 1830–1837), Sir J. E. Smith (404, 1789–1828), C. Sutton (11, 1797–1805), G. J. Symons (11, 1872–1888), R. Taylor (26, 1801–1842), Duke and Duchess of Teck (106, 1875–1908), J. Templeton (13, 1797–1822), W. Todhunter (27, 1806–1830), D. Turner (96, 1794–1830), T. Walford (14, 1795–1815), N. T. Wetherell (23, 1832–1871) and N. J. Winch (17, 1801–1829).

One box: letters from members of the family, including P. E. Barnes (28,

1814–1869), Miss C. A. Sowerby (21, 1817–1824) and Miss F. E. A. Sowerby (11, 1824–1838).

Three boxes of Mss: papers relating to books produced jointly by two or more members of the family, including *Flora Graeca, Mineral Conchology, Zoological Journal* and *Genera of Recent and Fossil Shells*, 1806–1835.

Three boxes of Mss: bills and receipts from tradesmen, suppliers, printers and others, 1782–1899.
L Sowerby Coll A

(2) Six boxes of Mss: correspondence and papers of James Sowerby, 1775–1822, including letters from J. de C. Sowerby (29, 1802–1822) and G. B. Sowerby I (19, 1803–1819); Mss, plates and drawings relating to *Flora Londinensis, The Pomona Britanica, Flora Luxurians, Medical Botany, British Mineralogy, British Miscellany* and other works; Mss relating to dry rot in the Royal Navy; unidentified botanical and zoological plates and drawings.
L Sowerby Coll B

(3) Seven boxes of Mss: correspondence and papers of J. de C. Sowerby, 1797–1871, including letters from J. C. Edwards (25, 1814–1834), T. Edwards (10, 1813–1840), C. E. Sowerby (83, 1824–1840), G. B. Sowerby I (41, 1812–1840) and J. Sowerby jun. (15, 1827–1833); Mss, plates and drawings relating to papers in *Transactions of the Linnean Society* and *Annals and Magazine of Natural History, Encyclopaedia of Plants* (1829) by J. C. Loudon, *Monograph of the Testudinata* (1836–1842) by T. Bell, *Silurian System* (1839) by R. I. Murchison, *Geology and Fossils of Sussex* (1850) by F. Dixon and other works.
L Sowerby Coll C

(4) Four boxes of Mss: correspondence and papers of the descendants of James de Carle Sowerby, principally William and James Bryant Sowerby, 1828–1955, including William Sowerby's diaries and account books (21 volumes, 1869–1899); letters from Mrs C. F. Sowerby (23, 1871–1879); drawings and plates for *History of British Zoophytes* (1838) by G. Johnston, *Figures and Descriptions of Canadian Organic Remains* (1859) by J. W. Salter, and *British*

Conchology (1862–1869) by J. G. Jeffreys; letters on the freshwater medusa discovered by William Sowerby in 1880.
L Sowerby Coll D

(5) Two boxes of Mss: correspondence and papers of James Sowerby's other children: George Brettingham, Charles Edward, Charlotte, Felicia, and their descendants, 1809–1886; including letters from James E. Bryant to his parents (183, 1852–1872); drawings for *Conchological Illustrations* (1832–1841) by G. B. Sowerby II, and *British Wild Flowers* (1858–1860) by John E. Sowerby.
L Sowerby Coll E, F, G, H

(6) One box of Mss: papers relating to the Royal Botanic Society, 1838–1922, including official letters of James de Carle, William and James Bryant Sowerby, successive secretaries.
L Sowerby Coll J

(7) Four boxes of Mss: unidentified drawings and plates by members of the Sowerby family and others.
L Sowerby Coll K

(8) Three boxes of metal plates: engraved copper and steel plates, 1788–1821, produced for *A Botanical Drawing Book* (1789), *British Miscellany*, *Monograph of the Eocene Cephalopoda* (1849–1860) by Edwards and Wood, *Monograph of Fossil Lepadidae* (1851) by C. Darwin, *Figures and Descriptions of Canadian Organic Remains* by J. W. Salter, and other publications.
L Sowerby Coll L

(9) Three boxes of Mss: the papers of J. W. Salter, 1813–1900, including autobiographical notes, 1852–1865; three scrapbooks of prints and drawings, 1813–1856; landscape, figure and natural history drawings by his wife and daughter, 1834–1875.
L Sowerby Coll N

(10) One scrapbook of plates and drawings of recent and fossil shells by J. Sowerby and others, made up by Miss E. Benett, 1819–1840.
L Sowerby Coll O12

107. **STONE, Sarah (1761/2–1844)**

Sarah Stone was born in London and lived there all her life. She showed a talent for drawing very early in life, and by the age of 20 was drawing plants, fossils and animals, especially newly imported species, for patrons. She worked in Sir Ashton Lever's Museum, using mounted specimens brought from all over the world. In 1790 her drawings of Australian animals were published in John White's *Journal of a Voyage to New South Wales* (1790). She married John Langdale Smith in 1790, after which her professional output declined, although she continued painting for her own pleasure.

Biographical Reference: personal communication from Christine Jackson.

Collection

(1) Nine boxes: 93 watercolour drawings of mammals, birds, fossils, corals and shells, painted from specimens in Sir Ashton Lever's Museum, 1781–1785.
L MSS STO

(2) One folder: 13 watercolour drawings of birds, [c.1788]. Bequeathed by Lord Rothschild, 1937.
L MSS STO

108. **STRICKLAND, Catherine Dorcas Maule**

The second daughter of Sir William Jardine (q.v.), who married Hugh Edwin Strickland (q.v.) in 1845.

Biographical Reference: *DNB*, (for her husband).

Collection

(1) One sketchbook: pencil sketches of landscape in England and Scotland, church architecture, antiquities and fossils, with photographs of Deerhurst Church, Gloucestershire, 1855–1864.
L MSS STR

109. **STRICKLAND, Hugh Edwin (1811–1853)**

Born in Yorkshire and educated at Rugby and Oxford, Strickland became interested in geology after attending the lectures of William Buckland (q.v.). He toured Asia Minor with W. J. Hamilton in 1835, studied the New Red Sandstone in central England in the late 1830s, and became involved in the reform of zoological nomenclature in 1840–1842. He toured Europe after his marriage in 1845, and thereafter turned his attention to ornithology, under the influence of his father-in-law, Sir William Jardine (q.v.). Strickland was accidentally killed while making a geological survey of railway cuttings in 1853.

Biographical Reference: *DNB*.

Collection

(1) One Ms volume: journal of a visit to the Continent by Hugh Strickland and his wife Catherine (q.v.), 1845, following their marriage. Includes sketches and printed ephemera. Presented by Dr N. B. Kinnear, 1948.
L MSS STR

110. **SYKES, William Henry (1790–1872)**

Born in Yorkshire, Sykes joined the army of the East India Company as a Cadet in 1803 and served in India until he retired with the rank of Colonel in 1833. Sykes acted as an interpreter, and became the Statistical Reporter, responsible for collecting information on the population and natural history of the Deccan region of central India. He returned to England in 1833, was made Chairman of the East India Company in 1856, and was Member of Parliament for Aberdeen from 1857 until his death.

Biographical Reference: *DNB*.

Collection

(1) Twenty Ms volumes: *Agriculture*, being descriptions and observations on plants, animals and agricultural implements of the Deccan, illustrated with

over 280 watercolour drawings by native artists, 1825–1830. The first volume contains a few of Sykes's own pencil sketches.

One Ms volume: *Reports on the Dakhin,* covering the mammals, birds, fish, reptiles and amphibia of the Deccan, together with meteorological observations, 1831, illustrated with watercolour drawings by native artists. Purchased 1920.
L MSS SYK

111. **TAMS, Willie Horace Thomas (1891–1980)**

Born in Cambridge, Tams developed an early interest in natural history, and worked as an assistant to Professor J. S. Gardiner (1872–1946) in the Zoology Department. He joined the staff of the Linnean Society in 1913. Tams served in the Canadian Army during the First World War, and was appointed Assistant Keeper in the Entomology Department of the Museum in 1920. He was a keen lepidopterist, and published over seventy papers on Lepidoptera taxonomy, including several on the Lasiocampidae.

Biographical Reference: 1981. *Proceedings of the British Entomological and Natural History Society* 14:34–40.

Collection

(1) Three letters in a folder: two letters regarding the offer of the OBE to Tams in the Coronation honours list, with reply, 1953.
L MSS TAM

112. **'TERRA NOVA'**

A steam barque, built in 1884 in Dundee. Hired by the Admiralty in 1903 to relieve Captain Scott in the Antarctic, *Terra Nova* was the ship used for Scott's British Antarctic Expedition, 1910–1913. She was sunk by enemy action off Newfoundland in 1943.

Historical Reference: Rice, A. L. 1986. *British Oceanographic Vessels, 1800–1950*. 193pp. London: Ray Society.

Collection

(1) One volume: *Night Order Book, British Antarctic Expedition,* 1910. The volume is mostly blank, but contains a list of stores needed for the expedition.
L MSS TER

113. **THOMPSON, Arthur Robert (1890–1939)**

Born in India and educated at Walthamstow Grammar School, Thompson was in business in the City of London as a young man. He served in the Royal Naval Volunteer Reserve during the First World War, and later became an Inspector of Taxes for the Inland Revenue. Throughout his life Thompson was an avid observer of the animals and plants of the English countryside, backing up his detailed notes with high-quality photographs. He published two books: *Nature by Night* (1931) and *Nature by Day* (1932).

Biographical Reference: 1940. *Proceedings of the Linnean Society, 152nd session*:375–377.

Collection

(1) Seventeen Ms volumes: natural history diaries (five volumes), notes and descriptions of British vertebrates (three volumes), observations on animals and plants (five volumes), butterfly and moth diaries (two volumes), albums of photographs and magazine cuttings (two volumes), 1908–1939. A finding aid is available. Presented by Mr E. P. Thompson, 1985.
L MSS THO

114. **TONGE, Mrs Olivia Frances (1858–1949)**

Olivia Frances Fitzmaurice was born in Glamorganshire, and was an accomplished watercolour painter from an early age, specializing in flowers, birds and reptiles. After her marriage in 1877 she took up woodcarving, metalwork, singing, dress designing and embroidery. She made two trips to India in 1908–1910 and 1912–1913, when she produced the notebooks in the collection.

Biographical Reference: information from the donor.

114(1) Tonge, Mrs Olivia Frances (1858–1949)
A watercolour drawing of geckos from Sketchbook 16 of the Tonge Collection. All the drawings are carefully annotated, although generally not dated.

Collection

(1) Sixteen sketchbooks: annotated watercolour drawings of flowers, fruit, mammals, birds, fish, crabs, insects, local jewellery, musical instruments, methods of transport etc., from Sindh, Calcutta, Darjeeling and elsewhere in India, 1908–1913. Presented by H. F. Hobbs, 1952.
L MSS TON

115. **TURNER, Dawson (1755–1858)**

Turner was born in Norfolk, and educated at North Walsham and Cambridge University. He took charge of the family bank in 1777 on the death of his father, and remained a banker all his life. Turner was interested in all aspects of natural history, but was principally a botanist, studying the lichens, fungi and other cryptogams. He was a collector of books and manuscripts, and built up a magnificent library, which was sold at auction after his death.

Biographical Reference: *DNB*.

Collection

(1) One volume of Mss: 13 printed and Ms items relating to the strata and fossils of Norfolk by Joseph Arnold, William Smith, J. de C. Sowerby, Richard Taylor and Dawson Turner, 1812–1825. A finding aid is available. Collected by Dawson Turner, and sold from his library in 1859; purchased by Warren R. Dawson in 1934, and presented by him to the Museum, 1958.
L MSS TUR

116. **TYLOR, Mrs Anna**

Anna Fox, the daughter of Sylvanus Fox of Cornwall, married Edward Burnett Tylor (1832–1917) in 1858. He was an archaeologist and anthropologist, who became the first Professor of Anthropology at Oxford University. He was author of *Primitive Culture* (1871).

Biographical Reference: *DNB* (for her husband).

Collection

(1) One notebook: a chronicle of the life of Sir Edward Burnett Tylor from 1857 to 1917, together with two letters to his brother, Alfred Tylor, 1875. Purchased 1969.
L MSS TYL

117. **VICTORIA COUNTY HISTORY**

The Victoria County History was founded in 1899 as a national historic survey which would form a worthy and permanent memorial to Queen Victoria. The series of publications was dedicated, by permission, to the Queen. Early issues included descriptions of the geology, botany and zoology of the counties, but these were dropped in volumes published after the Second World War.

Historical Reference: Pugh, R. B. 1967. Structure and aims of the Victoria History of the Counties of England. *Bulletin of the Institute of Historical Research* 40:65–73.

Collection

(1) Twelve boxed volumes of Mss, Tss and printed proofs: unpublished descriptions of the botany, geology, palaeontology, and vertebrate and invertebrate zoology of Cheshire, Dorset, Lincolnshire, Westmorland and Wiltshire, by O. P. Cambridge, C. W. Dale, J. A. Martindale, T. R. R. Stebbing and others, 1900–1908. A finding aid is available. Deposited on indefinite loan by the Victoria County History Committee, 1965.
L MSS VIC

118. **WALCOTT, John (c.1750–1831)**

Born in Creagh near Rathkeale, Co. Limerick, the son of a country gentleman, Walcott lived in Bath during his early years. He became an industrious and wide-ranging naturalist, who wrote books on British plants, the fossils found around Bath and *Synopsis of British Birds* (1789–1792). He lived in Teignmouth, Bristol, London and Southampton at different times, before returning to Bath in 1829.

Biographical Reference: Jackson, C. E. 1985. *Bird Etchings: the Illustrators and their Books, 1655–1855*. 292pp. London, pp. 148–156.

Collection

(1) Four volumes of drawings: small pen-and-ink drawings of fish, butterflies, flowering plants, ferns, mammals, reptiles, crustacea and other invertebrates, c.1770–1830. Some drawings are captioned. Most are by John Walcott, but at least one, dated 1833, is by his son W. H. L. Walcott. Purchased 1974.
L MSS WALC

119. **WALLACE, Alfred Russel (1823–1913)**

Wallace went to school in Hertford, England, before working as a surveyor and later a schoolteacher. He became interested in natural history and decided to try to earn his living by organizing collecting expeditions. Wallace travelled to South America in 1848 and spent the years 1854 to 1862 in what are now Indonesia and Malaysia. He became interested in the problem of the origin of species, and arrived at the idea of natural selection at the same time as, and independently from, Charles Darwin (q.v.). Wallace made important contributions to biogeography, and the boundary between the Indo-Malayan and Australian regions is still called the Wallace Line.

Biographical Reference: *DSB*.

Collection

(1) Letters in a folder: four letters to Lord Walden with one draft reply, 1871–1872; one letter to G. A. Boulenger, 1904, and one to C. O. Waterhouse, 1908. L MSS WALL

120. **WALLICH, George Charles (1815–1899)**

Born in Calcutta, the eldest son of Nathaniel Wallich (1786–1854) the botanist, George Wallich was educated in England. He joined the Indian Army as a surgeon in 1838, and saw service in Bengal and the Punjab before being invalided out in 1857 with the rank of Surgeon-Major. Wallich became an authority on the structure and distribution of the Protozoa, and was appointed naturalist to HMS *Bulldog* on a surveying expedition to the North Atlantic Ocean in 1860. He published books and papers as a result of the voyage, and engaged in a series of controversies over the significance of his observations.

Biographical Reference: *DSB*.

Collections

(1) Three Mss in a folder: *On the Flint Veins, or Fissure Flints of the Upper Chalk*, undated; notes and drawings giving results of dredging from HMS

Bulldog off the west coast of Greenland, [1860]; copies of letters from Sir Charles Lyell to Wallich, undated.
L MSS WAL A

(2) Three Ms volumes: *Report on the Soundings Taken on Board HM Ship Bulldog*, 1860; *Private Journal No. 1*, June to October 1860; *Private Journal No. 2*, October to November 1860. Purchased 1986.
L MSS WAL B

(3) Four boxes of letters: scientific correspondence, 195 letters from 60 correspondents, 1851–1898, including 8th Duke of Argyll (15, 1893–1894), W. B. Carpenter (10, 1865–1889), and W. B. Turner (48, 1884–1898); non-scientific correspondence, 186 letters from 24 correspondents, 1848–1899, including G. Blake (25, 1861–1868), E. Norton (32, 1855–1858), N. D. S. Wallich (15, 1865–1862), S. M. Wallich (11, 1859–1870) and J. Wilkinson (41, 1858–1859); with other Mss and printed papers relating to Wallich's service in the army, his medical work, his inventions, financial affairs; diaries for 1860 and 1861; and *Eminent Men of the Day Photographed by G. C. Wallich*, 1870. A finding aid is available. Purchased 1985.
L MSS WAL C

121. **WATERHOUSE, Alfred (1830–1905)**

Alfred Waterhouse was born in Liverpool. After leaving school he was articled to an architectural practice in Manchester. On finding the work to his liking, he started his own practice in the city in 1853 and established a large clientele in the north of England. Waterhouse moved to London in 1865, and there his reputation attracted much work from the southern counties. His disappointment when he just failed to secure the commission for the new London law courts in 1866 was short-lived, for in 1868 he was awarded the contract for creating a Natural History Museum in South Kensington for the government. The work was completed in 1880. The plan is broad and simple, yet the architecture is marked by great richness of decoration to the terracotta brickwork.

Biographical Reference: *DNB*.,

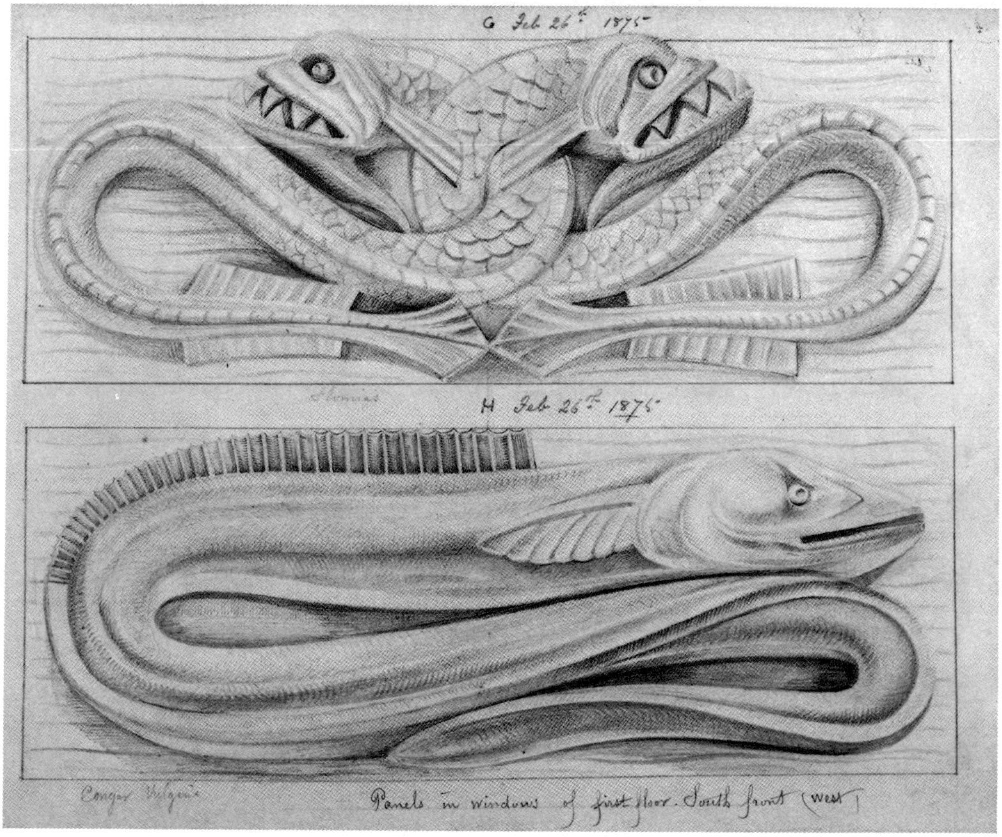

121(1) Waterhouse, Alfred (1830–1905)
Pencil drawing for a terracotta panel for The Natural History Museum, depicting the oceanic fish *Stomias* and a conger eel, dated 26 February 1875.

Collection

(1) Seven boxes of drawings: 134 pencil drawings of plants, animals and fossils for the architectural decoration of The Natural History Museum, [c.1874–1879]. Presented by Mr Michael Waterhouse, 1962.
L MSS WAT

122. **WATERHOUSE, George Robert (1810–1888)**

Born in London and educated in Belgium, Waterhouse worked as an architect as a young man. He was interested in entomology, and was one of the founders of the Entomological Society of London in 1833. He worked at the Royal

Institution, Liverpool, before becoming Curator to the Zoological Society of London in 1836, and an Assistant in the Mineralogical Branch of the British Museum in 1843. Waterhouse became Keeper of the Mineralogical and Geological Branch in 1851, and then Keeper of Geology when the departments were reorganized in 1857. He retired in 1880. In spite of his posts at the Museum, Waterhouse's main interests were in the Coleoptera and Mammalia.

Biographical Reference: *DNB*.

Collections

(1) One folder: five letters to his father, J. E. Waterhouse, 1836–1838.
L MSS WAT A

(2) One folder of Mss: letters and memoranda on his employment with the Zoological Society and British Museum, 1839–1871; notes on mythological and other subjects; 23 letters from 19 correspondents on zoological topics. A finding aid is available.
L MSS WAT B

123. **WATERTON, Charles (1782–1865)**

Charles Waterton was born in Yorkshire of a distinguished family of landowners. He was educated in Durham and Lancashire, and visited Spain while still a young man. He was a great traveller, living in British Guiana from 1804–1813, and revisiting in 1816 and 1820. In 1824 he travelled in North America, the West Indies and British Guiana. Waterton wrote the very popular *Wanderings in South America* in 1825. He succeeded to the family estate of Walton Hall in 1806, and developed the grounds as a nature reserve and a home for the exotic animals that he brought back from overseas.

Biographical Reference: *DSB*.

Collection

(1) Letter in a folder: letter to Mr Banks, 1849, making an appointment to visit. Purchased 1975.
L MSS WATER

124. **WHITE, Errol Ivor (1901–1985)**

White was born and educated in London, gaining a B.Sc. in geology from King's College, London, in 1921. In 1922 he joined the Geology Department of The Natural History Museum, working in the Fish Section. He devoted his working life to fossil fish, collecting with Museum expeditions in Madagascar and Spitzbergen, and specializing in the primitive fish of the Devonian Period. White became Deputy Keeper in 1938 and Keeper of Geology in 1955. He was elected FRS in 1956 and President of the Linnean Society in 1964. White retired in 1966.

Biographical Reference: *DNB*.

Collection

(1) Five photograph albums: two containing photographs of Scotland, France, The Natural History Museum and London Zoo, 1928; three covering the Vernay – Archbold Expedition to Madagascar, 1929, including also photographs taken in Zanzibar, Kenya and Aden. Presented by Abingdon Police, 1989.
L MSS WHI

125. **WHITEHEAD, Peter James Palmer (1930–1992)**

Born in Kenya and educated at Cambridge, Whitehead studied freshwater fish in East Africa from 1953 until 1960, when he joined the Zoology Department of The Natural History Museum. He worked in the Fish Section, making a particular study of the clupeoids (herrings), and published *Clupeoid Fishes of the World* (two volumes, 1985–1989). Whitehead's taxonomic research led him to the early literature of ichthyology, and he became interested in the history of zoology. He was an artist and writer as well as a scientist.

Biographical Reference: The *Independent*, 29 May 1992.

Collection

(1) One Ts in a folder: *To Brazil with Count Maurits*, the text of a lecture delivered to the Osler Club, 1973. Presented by the author, 1973.
L MSS WHITE

126. **WILKINS, Sir (George) Hubert (1888–1958)**

Wilkins was born in South Australia and studied at Adelaide School of Mines from 1903 to 1908. Thereafter he had a varied career as a journalist, aviator, photographer and explorer. He travelled in the Canadian Arctic (1913), the Antarctic (1920–1921) and in Northern Australia for The Natural History Museum (1923–1925). Wilkins published an account of this expedition entitled *Undiscovered Australia* in 1928. He made pioneering flights over the Arctic and Antarctic, and travelled under the polar ice cap in a submarine.

Biographical Reference: *DNB*.

Collections

(1) Three volumes of Mss and photographs: two Ts drafts of *Undiscovered Australia*, 1925, with 87 photographs taken in Northern Australia, 1923–1925.
L MSS WIL A

(2) Two boxes of photographs: 113 copy prints and 40 copy glass negatives taken in Northern Australia, 1923–1925, some published in *Undiscovered Australia* (1928).
L MSS WIL B

127. **WINTLE, William James (1861–1934)**

Born in Gloucestershire and educated in London, Wintle worked first as a schoolteacher, before turning to writing and journalism. He was a religious and charitable man, who lived for a time among the monks of Caldy Island, South Wales. Wintle was a wide-ranging naturalist with a particular interest in the land and freshwater molluscs of the British Isles. He wrote *Recreations with a Pocket Lens* (1911).

Biographical Reference: 1934, Obituary. *Proceedings of the Malacological Society* 21:149.

Collection

(1) One volume of letters and photographs: 33 letters to B. B. Woodward on the natural history of Caldy Island, with a map, postcards and photographs, 1920–1922. Presented by H. Clarke, 1938.
L MSS WIN

128. **WOOD, Casey Albert (1856–1942)**

Casey Wood was born in Canada of American parents, and educated in Ottawa and at McGill University. He trained as an ophthalmologist and practised in London and Chicago. Wood was a keen ornithologist, and retired in 1907 to devote himself to birds. He travelled widely in the Far East, visiting India, Ceylon and the Pacific. Wood was a great book collector and founded libraries of ophthalmology and ornithology at McGill University. He wrote *An Introduction to the Literature of Vertebrate Zoology* (1931).

Biographical Reference: 1942. Obituary. *Ibis*, 14th series, 6:528–530.

Collection

(1) One bound Ts: a duplicated letter to his friends describing the natural history of the West Indies and some parts of South America, 1922. Presented by the author, 1922.
L MSS WOO

129. **WOODWARD, Bernard Bolingbroke (1816–1869)**

Born and schooled in Norwich, Bernard was the eldest son of Samuel, and the brother of Samuel Pickworth Woodward (q.v.). He continued his studies in London, and became a minister in the Congregational Church in 1843.

Woodward had wide interests in natural history and antiquities, and was appointed Librarian in Ordinary to the Queen at Windsor Castle in 1860.

Biographical Reference: *DNB*.

Collection

(1) Four Ms volumes: *Notes of a Naturalist* and *Notes from Nature*, a natural history diary, mainly botanical, consisting of observations made around Norwich, with sketches, 1832–1840. The diary includes contributions by S. P. Woodward, and transcripts from works by W. J. Bree and others. Presented by Bernard Barham Woodward, 1914.
L MSS WOOD

130. **WOODWARD, Samuel Pickworth (1821–1865)**

Born and educated in Norwich, Samuel Pickworth was the second son of Samuel, and the brother of Bernard Bolingbroke Woodward (q.v.). He studied natural history and geology, and worked for a time as an assistant to Dawson Turner (q.v.). He was appointed Sub-Curator to the Geological Society of London in 1839, and Professor of Geology at the Royal Agricultural College in Cirencester in 1845. In 1848 Woodward obtained the post of Assistant in the Mineralogical and Geological Branch of the British Museum. His most popular work was *Manual of the Mollusca* (1851–1856).

Biographical Reference: *DNB*.

Collection

(1) One Ms volume: *Poetic Gleanings*, a commonplace book of poetry and prose, much of it related to natural history, undated. Presented by C. Kirke Swan, 1974.
L MSS WOODW

131. **YONGE, Sir (Charles) Maurice (1899–1986)**

Yonge went to school in Wakefield and read zoology at Edinburgh University, where he was the Baxter Natural Science Scholar. Work at the Marine Biological Association, Plymouth, was followed by Yonge's appointment as leader of the Great Barrier Reef Expedition, 1928–1929. Thereafter he was Professor of Zoology at Bristol, 1933–1944, and Glasgow, 1944–1964. Yonge's research was almost entirely concerned with molluscs and coral reef ecology. He was elected FRS in 1946 and was awarded the Darwin Medal of the Royal Society in 1968.

Biographical Reference: 1989. Obituary. *The Yearbook of the Royal Society of Edinburgh, 1989*. 187pp. Edinburgh. pp. 73–75.

Collection

(1) Forty-four boxes of Mss and other materials: the papers of Sir Maurice Yonge, 1922–1982. Presented by Lady Yonge, 1990.
L MSS YON

The collection was catalogued by the National Cataloguing Unit for the Archives of Contemporary Scientists in 1991, and is arranged in six divisions:

A. Diaries. Glasgow University pocket diaries (1952–1953), and Edinburgh University pocket diaries (1970–1983).

B. Notebooks. Relate to the Great Barrier Reef Expedition (1927–1929), and visits to the United States (1939 and 1956), Africa (1953 and 1962), Caroline Islands (1965) and Australia (1967).

C. Research. A large collection of research files relating to all aspects of Yonge's marine biological work (1922–1982), together with original drawings for his publications and card bibliographies. The files include correspondence, literature notes, field observations, experimental data, text drafts, photographs and drawings.

D. Societies and Organizations. Correspondence and printed papers document Yonge's membership and commitment to a number of scientific societies, including the Royal Society and the Royal Society of Edinburgh (1926–1981).

E. Correspondence. General correspondence with scientific colleagues arranged chronologically (1925–1989), together with groups of letters on the Great Barrier Reef Expedition, 'Miscellaneous Hong Kong and Pacific Science' and *Advances in Marine Biology*.

F. Non-print Material. Photographs relating to the Great Barrier Reef Expedition (1927–1929), and other photographs taken in the West Indies, East Africa, France, Italy and elsewhere (1930–1975).

Addenda: Photocopies in General Library Manuscripts

132. **AGASSIZ, Jean Louis Rodolphe (1807–1873)**

Seven letters, 1828–1829, in Harvard University Library.
L MSS AGA B

133. **BAKER, Henry (1698–1771)**

Correspondence with William Arderon, 1744–1767, in the Dyce Collection, Victoria and Albert Museum, London.
L MSS BAK

134. **BANKS, Sir Joseph (1743–1820)**

Reference to the plans and survey of Revesby in the County of Lincolnshire, by Thomas Stone, 1794, in Lincolnshire County Record Office.
L MSS BAN B

A survey of all the . . . freehold lands and possessions of William Banks, by Thomas Fowler, 1754, in Lincolnshire County Record Office.
L MSS BAN B

[Estate accounts of William Banks], 1750, in Lincolnshire County Record Office.
L MSS BAN B

Wills of Sir Joseph and Lady Dorothea Banks, 1820 and 1822, in the Public Record Office, London.
L MSS BAN C

Inventory of Sir J. Banks's library as received by the British Museum, 1827, in the British Library.
L MSS BAN D

135. **BANKS, Sir Thomas Christopher (1765–1854)**

Transcripts of letters to John Banks on the Banks family, 1853–1854, in private hands.
L MSS BANK

136. **BLEEKER, Pieter (1819–1878)**

Draft of a letter to Dr Albert Gunther, 16 November 1867, in Leiden Museum.
L MSS BLE

137. **BURGIS, Mary J.**

An Ecological Study of the Zooplankton in Lake George, Uganda, a thesis submitted to the University of London for the degree of Ph.D., 1971.
L MSS BUR

138. **CARTER, Harold Burnell (b.1910)**

A First Account of the Spanish Merino Flock of His Majesty George III, with letters and a copy of a photograph, 1963–1964.
L MSS CART

The Sheep and Wool Correspondence of Sir Joseph Banks, the original Ts, [c.1972].
L MSS CART

139. **CLIFT, William (1775–1849)**

Copy of Dr Shaw's Catalogue of Part of the Natural History Preserved in Spirit in the Hunterian Museum, 1806, in the Royal College of Surgeons, London.
L MSS CLI

140. **COUCH, Jonathan (1789–1870)**

A Journal of Natural History, being the Result of my own Observations or Derived from Living Testimony, 1862–1866, in Cheltenham Public Library.
L MSS COU

141. **CRANCH, John (1785–1816)**

Remarks of Animals &c as were Observed During a Voyage of Discovery in the Congo Expedition with a List of Such Specimens as were Taken & Preserved, 1816, with a diary and drawings from the same expedition, in the Muséum National d'Histoire Naturelle, Paris.
L MSS CRA

142. **DARWIN, Charles Robert (1809–1882)**

[Ornithological notes], [?1834–1837], in Cambridge University Library (Mss 29(ii)).
L MSS DAR B

[Books read and to be read], four notebooks, 1838–1860, in Cambridge University Library (Mss 119, 120, 122, 128).
L MSS DAR B

[Four notebooks on the transmutation of species], 1837–1839, in Cambridge University Library (Mss 121–124).
L MSS DAR B

Excised pages from the transmutation notebooks, 1837–1839, in Cambridge University Library.
L MSS DAR B

Two notebooks on metaphysics, 1838, in Cambridge University Library (Mss 125–126).
L MSS DAR B

Miscellaneous Mss, various dates, in Cambridge University Library (Mss 5, 42, 45, 47, 49, 53, 76–78, 83, 88, 91, 109, 126).
L MSS DAR B

Questions about the Breeding of Animals, c.1840, annotated copy in Cambridge University Library.
L MSS DAR B

Darwin's *Beagle* notebooks, 1, 2 and A, 1835–1839, in Down House, Downe, Kent.
L MSS DAR C

The red notebook, 1836–1837, in Down House, Downe, Kent.
L MSS DAR C

The Journal of a Voyage in HMS Beagle (1979 – . A facsimile reprint of the Ms in Down House, Downe, Kent.
L MSS DAR C

Letters written by Charles Darwin to various correspondents, various dates, in the Department of Manuscripts, British Library.
L MSS DAR D

Letters written by Charles Darwin, together with letters about him, photographs and related printed ephemera, ?1846–1891, in the collection of Robert M. Stecher, Cleveland, Ohio.
L MSS DAR E

One letter to W. B. Tegetmeier, 27 Nov 1858, offered for sale by Scriptorium in 1982.
L MSS DAR F

Four letters from Darwin to F. D'A. Furtado, 1881, in Universidade de Lisboa, Lisbon.
L MSS DAR F

143. **DAY, Francis (1829–1889)**

Letters from Francis Day and Arthur O'Shaughnessy to Wilhelm Peeters, and from Day to Eduard von Martens and Franz Hilgendorf, 1872–1887, in Zoologisches Museum, Berlin; one letter from Francis Day to Leon Vaillant, 1875, in Muséum National d'Histoire Naturelle, Paris; journal, notebook, reports, letters, photographs and family trees relating to Francis Day and his

family, 1866–1976, in possession of the Egerton family; letters from Francis Day to the Under Secretary of State for India, 1865–1866, in the British Library (India Office Records); copies of wills, marriage and birth certificates of members of the Day family, obituaries of Francis Day, and other printed materials, 1863–1966.
L MSS DAY

144. **DUNN, Ian George**

Ecological Studies on the Fish of Lake George, Uganda; with Particular Reference to the Cichlid Genus Haplochromis, a thesis submitted to the University of London for the degree of Ph.D., 1972.
L MSS DUN

145. **FORSTER, Johann Georg Adam** (1754–1794)

Observationes Historiam Naturalem Ipectantes quas in Navigationes ad Terras Australes Instituere Coepit G. F., July 1772–May 1773, in Muséum National d'Histoire Naturelle, Paris.
L MSS FORS

146. **FORSTER, Johann Rheinhold** (1729–1798)

Descriptiones Animalium Suae in Itinire ad Maris Australis Terras Suscepto, Collegit, Descripsit & Delineavit Joannes Reinoldus Forster, four volumes, 1772–1775, in the Staatsbibliothek Preussischer Kulturbesitz (Ms Lat quart 133–136).
L MSS FORST A

Typed transcript of *Journal of a Journey from London to Plymouth and a Voyage on Board His Majesty's Ship the Resolution, Captain Cook Commander*, six volumes, 1772–1775, in the Staatsbibliothek Preussischer Kulturbesitz (Ms Germ quart 222–227).
L MSS FORST B

Five letters from Forster to unknown recipients, 1780–1786, in Staatsbibliothek Preussischer Kulturbesitz.
L MSS FORST C

147. **GODMAN, Frederick Ducane (1834–1919)**

The Early Partnerships of Whitbread and Company, Ts by Nicholas B. Redman, undated, supplied by The Whitbread Archive. Contains information on F. D. Godman and his family.
L MSS GOD

148. **GOODMAN, Susan Louise (b.1946)**

Illustration and Iconography: Representations of the Natural World before 1700, Ts lecture delivered in the Museum's 'Art in Natural History' series, 1991.
L MSS GOO

149. **GORDON, Robert Jacob (1741–1795)**

Notes on the animals, plants and minerals of the Cape of Good Hope, correspondence, diaries and logbooks of his voyages, c.1775–1797, in Stafford County Record Office (Duke of Sutherland Mss).
L MSS GOR

150. **GWAHABA, James Joshua**

Population Studies of the More Abundant Fish Species in Lake George, Uganda, a thesis submitted for the M.Sc. at Makerere University, Kampala, 1973.
L MSS GWA

151. **HOARE, Michael Edward (b.1941)**

The Contribution of Johann Reinhold and George Forster to the Literature of Cook's Second Voyage (1772–1775) . . ., a thesis submitted for the MA at Monash University, Clayton, VIctoria, Australia, 1970.
L MSS HOA

152. **LAET, Johannes de (1593–1649)**

De Plantis Brasiliensibus, c.1645, in the British Library (Sloane Ms 1554).
L MSS LAE

153. **LAISHLEY, Richard (1815–1897)**

Sketch Notes of Family History with Appendix and Journal of Voyage to England in 1883, 1890, in private hands.
L MSS LAI B

154. **LAYARD, Austen Henry (1817–1894)**

Eight letters from W. Buckland, J. D. Hooker and R. Owen, 1862–1869, in the American Philosophical Society Library, Philadelphia.
L MSS LAY

155. **MARCGRAVIUS, Georgius (1610–1644)**

Extracts from Marcgravius's astronomical notes, possibly by A-J Pingre, undated, in the Observatoire de Paris.
L MSS MAR

156. **MATY, Matthew (1718–1776)**

Biographical material relating to Maty, extracted and copied from a variety of contemporary sources in the British Museum Archives and elsewhere.
L MSS MATY

157. **MILLS, Eric L. (b.1936)**

Studies of the Clergyman and Naturalist Alfred Merle Norman (1831–1918), 1975. The photocopied text is illustrated with original photographs.
L MSS MIL

158. **MIVART, St George Jackson (1827–1900)**

Twenty-two letters from Mivart to Charles Darwin, 1868–1872, in Cambridge University Library.
L MSS MIV

159. **MURRAY, Andrew (1812–1878)**

Twenty letters from scientists, including Charles Darwin and T. H. Huxley, 1859–1878, in the collection of Dr Roger Pyrah.
L MSS MUR

160. **PALLAS, Peter Simon (1741–1811)**

A notebook/diary for 1762–1763, and eleven letters to and from Pallas, in the Staatsbibliothek, Dahlem, Germany.
L MSS PAL

161. **PENNANT, Thomas (1726–1798)**

Two lists of questions to be answered and projects to be undertaken by Joseph Banks on his voyage to Newfoundland, 1766, in the collection of the Earl of Denbigh, Pailton House, Warwickshire.
L MSS PEN

162. **PLAYFAIR, Sir Lyon (1818–1898)**

Index to the Playfair Correspondence Owned by Sir Edward Playfair (to be Donated to University of Dundee), 1972.
L MSS PLAY

163. **SALVIN, Osbert (1835–1898)**

Letters from Spencer F. Baird, 1860–1864; Charles Darwin, 1868–1871; Thomas L. Powys, Lord Lilford, 1871–1895; and Salvin to members of his family, 1858; in the collection of Sybil S. Rampen.
L MSS SALV

List of the Papers of Osbert Salvin in the Possession of Sybil S. Rampen, 1985.
L MSS SALV

164. SCHEUCHZER, Johann Jacob (1672–1733)

Letters from J. Petiver, 1704–1706; Sir Hans Sloane, 1706–1727; John Woodward, 1701–1722, and others, in the Zentralbibliothek, Zurich.
L MSS SCH

165. SHIRLEY, Ann M

Notes on Dr Scoresby's Whaling Logs/Journals, 1803, 1823–1824 and other Items in Whitby Literary & Philosophical Society's Collections, 1978.
L MSS SHI

166. SIEBOLD, Carl Theodore Ernst von (1804–1885)

Twelve letters to Theodor Whilhelm Engelmann, 1845–1851, in the Staatsbibliothek Preussischer Kulturbesitz, Berlin.
L MSS SIE

167. SLOANE, Sir Hans (1660–1753)

Letters to and from Sir Hans Sloane copied from originals in the Linnean Society, the Royal Society, the Wellcome Institute and a library in Göttingen; also an account of Sloane's life from the British Library (Ms Sl 4241).
L MSS SLO

168. VOYSEY, Henry Wesley (1791–1824)

Mr Voysey's Report to the Marquis of Hastings, Drawn up in June 1819, 1820, in the Bodleian Library, Oxford, Russell of Swallowfield papers; on the geology of India.
L MSS VOY

Report Presented to the Supreme Government dated 24 June 1820 Accompanied by a Map, Drawings and Specimens, 1820, in the Bodleian Library, Oxford, Russell of Swallowfield papers; on the geology of India.
L MSS VOY

169. **WALLACE, Alfred Russel (1823–1913)**

Transcripts of five letters to Samuel Stevens, 1856–1858, in private hands.
L MSS WALL

170. **WATLING, Thomas (1762–1806)**

Letters from an Exile at Botany Bay to his Aunt in Dumfries; Giving a Particular Account of the Settlement of New South Wales, with the Customs and Manners of the Inhabitants, [1793], in the British Library.
L MSS WATL

171. **WHITE, Gilbert (1720–1793)**

The Third Book of Tully de Oratore, 1741, in private hands.
L MSS WHIT

Account books and memoranda, 1758–1793, in the Gilbert White Museum, Selborne.
L MSS WHIT

Two volumes of Ms sermons on 1 Corinthians, 11, 25, and Hebrews 4, 14 (1718), in private hands.
L MSS WHIT

The Natural History of Selborne in the County of Southampton, 1788, White's original Ms, containing much unpublished material, in the Gilbert White Museum, Selborne.
L MSS WHIT

Index of Personal, Institutional and Ship Names

This index contains all names used as headings, and which appear in the catalogue entries, whether as authors, subjects or donors. Locations of originals of items in the photocopy collection are not indexed. References are to entry numbers, not pages.

Index of Places

This index contains place names relating to the subject matter and origin of the manuscripts and drawings, arranged alphabetically by broad geographical categories. Indexing goes down to the level of country, except for England, where counties are indexed where possible. Names used are normally those in use at the time the item was prepared.